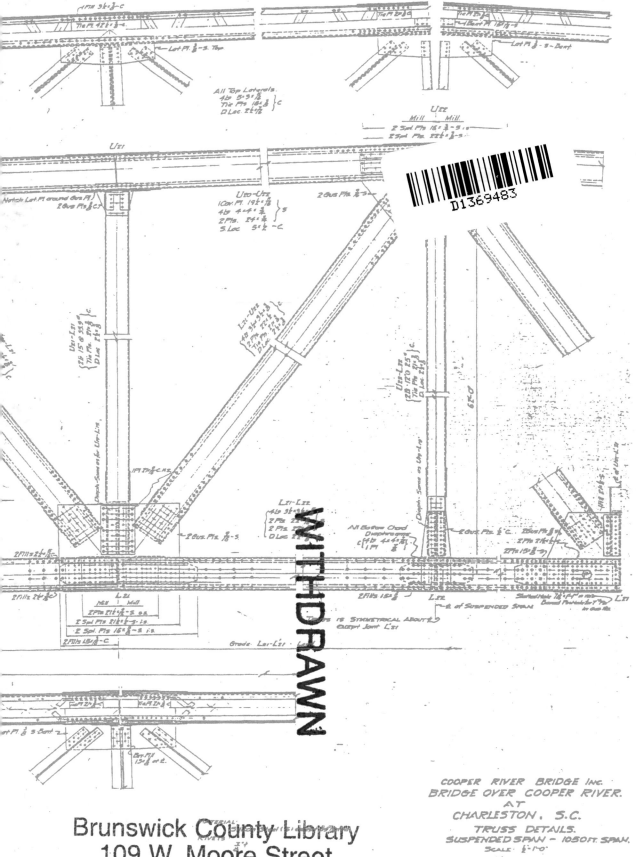

COOPER RIVER BRIDGE Inc.
BRIDGE OVER COOPER RIVER.
AT
CHARLESTON, S.C.
TRUSS DETAILS.
SUSPENDED SPAN - 1050 FT. SPAN.

WADDELL & HARDESTY
CONSULTING ENGINEERS
NEW YORK CITY
SHEET No 47
AUGUST 3, 1928

# *The Great*
# COOPER RIVER BRIDGE

JASON ANNAN *and* PAMELA GABRIEL

University of South Carolina Press

© 2002 University of South Carolina

Published in Columbia, South Carolina, by the
University of South Carolina Press

Manufactured in the United States of America

06  05  04  03      5  4  3  2

Library of Congress Cataloging-in-Publication Data

Annan, Jason, 1974–
   The great Cooper River Bridge / Jason Annan and Pamela Gabriel.
      p. cm.
   Includes bibliographical references (p. ) and index.
   ISBN 1-57003-470-2 (cloth : alk. paper)
   1. John P. Grace Memorial Bridge (Charleston, S.C.) I. Gabriel, Pamela,
1944–  . II. Title.
TG25.C36A56 2002
388.1'32'09757915—dc21                                        2002012312

*To Our Families*

# CONTENTS

Illustrations and Tables / ix

Acknowledgments / xiii

Introduction / xv

*Chapter One*

The Dream / 1

    The First Bridge: The Ashley River Bridge / 2
    The Coney Island of the South / 4
    Sullivan's Island / 12
    New Owners, New Dreams / 16

*Chapter Two*

Change Comes to Charleston / 22

    The Naval Base and Tourism / 23
    The Great Debate: A Private or Public Bridge / 32

*Chapter Three*

John Patrick Grace / 35

    Charleston Roots / 35
    Grace in Public Life / 37
    The Burden of a Lawyer / 40

*Chapter Four*

Building the Bridge / 44

    Selecting a Location / 44
    Waddell and Hardesty, Bridge Engineers / 46
    Selecting a Bridge Design / 48
    A Cantilevered Truss / 53
    The Bridge Is Built / 57
    Tragedy Strikes / 63
    Construction of the Superstructure / 67
    The Completed Bridge / 72

*Chapter Five*

Hope and Despair / 78

  The "Great Cooper River Bridge" Is Opened / 78
  The Hopes of the '20s, the Reality of the '30s / 88
  Purchase and Repurchase / 93
  The Bridge Is Free / 95

*Chapter Six*

The Bridge Comes of Age / 98

  The *Nicaragua Victory* and Tragedy / 99
  Growth of East Cooper / 102
  Mishaps and Memories / 108
  The New Cooper River Bridge / 111
  The Bridge Run / 113
  The End of the Bridges / 118

Epilogue / 125


Appendix / 127

Notes / 131

Bibliography / 139

Index / 141

# Illustrations and Tables

The Great Cooper River Bridge, Charleston, S.C. / xviii

1.1 Early map of Charleston and vicinity from *Harper's Magazine*, circa 1890 / 1

1.2 The Ashley River Bridge tollhouse as it appeared around 1900 / 2

1.3 Photograph, circa 1890, showing the wooden bridge across the Ashley River / 3

1.4 Postcard, circa, 1930, depicting the Ashley River Memorial Bridge / 5

1.5 The Santee River Bridge, located between McClellanville and Georgetown, S.C., completed in the mid-1920s / 5

1.6 Cove Inlet Bridge, between Mount Pleasant and Sullivan's Island, circa 1920 / 7

1.7 The trolley station on the Isle of Palms, circa 1910 / 7

1.8 Ferry map from early 1900s published by the Consolidated Railroad, Gas and Electric Company / 8

1.9 Electric trolley used by Consolidated Railroad, Gas and Electric Company to transport vacationers to the Isle of Palms during the early 1900s / 9

1.10 Postcard from the early 1900s showing the Hotel Seashore on the Isle of Palms / 10

1.11 The Ferris wheel at the Isle of Palms, shown in 1913 postcard / 10

1.12 The *Commodore*, a double-ended ferry and the largest ship in the ferry fleet / 11

1.13 Cover of early 1900s promotional brochure for the beach resort on Isle of Palms / 12

1.14 Illustration from an advertisement for the Isle of Palms that appeared in 1903 / 13

1.15 The village of Moultrieville on Sullivan's Island as it appeared circa 1850 / 15

1.16 Atlantic Beach Hotel on Sullivan's Island in the early 1900s / 16

1.17 Bathers on Sullivan's Island / 17

1.18 The *Nansemond*, one of several ferries that transported automobiles across the Cooper River during the 1920s / 18

1.19 Automobiles disembarking from one of the Cooper River ferries / 19

1.20 The hotel and pavilion on the Isle of Palms in the 1920s / 20

2.1 An impoverished area in Charleston in the early 1900s / 23

2.2 The old immigration station, Charleston, S.C. / 25

2.3  Postcard, circa 1920, showing the Peoples Building on Broad Street / 27

2.4  Postcard from the early 1900s showing the Fort Sumter Hotel / 27

2.5  The Charleston Hotel as it appeared in the early 1920s / 28

2.6  The Francis Marion Hotel, circa 1930 / 29

2.7  Abandoned gas station, built in the 1920s, at the corner of Ashley Avenue and Wentworth Drive / 29

2.8  King Street in the 1920s / 30

2.9  Postcard showing Charleston Harbor in the early 1900s / 31

2.10  Early Charleston tour book, circa 1920 / 32

3.1  Photograph of John Patrick Grace taken in 1929 / 36

3.2  John Grace's home at 174 Broad Street / 38

3.3  Memorial tablet to Mayor John Grace, located in Charleston City Hall / 41

3.4  John Grace's gravesite in St. Lawrence Cemetery, Charleston, S.C. / 43

4.1  Illustration from 1926 of the proposed Market Street approach to the Cooper River Bridge / 45

4.2  Dr. John Alexander Low Waddell as he appeared at the height of his career / 47

4.3  Shortridge Hardesty, designer of the Cooper River Bridge / 48

4.4  Table: Significant cantilevered truss bridges / 49

4.5  Page from the original blueprints for the Cooper River Bridge / 51

4.6  Firth of Forth Bridge in Scotland, the first steel bridge to use a cantilevered truss / 54

4.7  Photograph taken shortly before the completion of the Cooper River Bridge / 56

4.8  Photograph taken on June 18, 1929, showing progress on the Cooper River span of the Cooper River Bridge / 57

4.9  Photograph taken on May 23, 1929, showing one of the Cooper River Bridge's many bearing shoes / 58

4.10  Photograph taken on July 2, 1929, showing construction of the suspended span over the Cooper River / 59

4.11  Charles Keyes Allen / 61

4.12  Aftermath of the collapse of one of the pneumatic caissons, December 2, 1929 / 64

4.13  Photograph of the tilted caisson / 65

4.14  Construction of the Cooper River Bridge, May 29, 1929 / 66

4.15  Steel workers balance on 18-inch-wide beams used in the trusses of the Cooper River Bridge / 66

4.16  Training dummies, like the ones shown in this June 8, 1929 photograph, were used in safety exercises during construction of the Cooper River Bridge / 67

4.17  Town Creek span of the Cooper River Bridge, February 1, 1929 / 68

4.18  Construction of the anchor arm of the Town Creek span closest to the Charleston shoreline, December 28, 1928 / 68

4.19  Massive cranes used to lift heavy steel beams into position / 69

4.20  Large cranes on the ends of each cantilever arm of the Cooper River span, June 18, 1929 / 70

4.21  Progress on the Cooper River span, May 10, 1929 / 71

4.22  Temporary scaffolding, or falsework, supporting an incomplete anchor arm and crane, May 17, 1929 / 71

4.23  Workers laying one of the bearing shoes used to tie the bridge superstructure to the piers, May 17, 1929 / 72

4.24  Completion of the concrete road deck on the Cooper River Bridge, July 17, 1929 / 73

4.25  The Charleston approach to the Cooper River Bridge, at the old intersection of Lee and America Streets / 74

4.26  The incomplete bridge as it appeared from the top of the Cooper River span, March 27, 1929 / 75

4.27  Table: Cooper River Bridge statistics / 76

4.28  Aerial photograph taken on May 9, 1929, showing progress on the Cooper River Bridge / 77

5.1  Brochure for the opening of the Cooper River Bridge, August 8–10, 1929 / 79

5.2  Corporate logo of the Cooper River Bridge, Inc. / 80

5.3  Ribbon cutting to open the Cooper River Bridge, August 8, 1929 / 82

5.4  Paramount Studios filming the new Cooper River Bridge shortly before the bridge's official opening / 82

5.5  One of the floats in the opening parade for the Cooper River Bridge / 83

5.6  Historic Floats Parade, August 9, 1929 / 83

5.7  The Cooper River Bridge, an important link in the Coastal Highway / 84

5.8  "Ode to the Cooper River Bridge" by Furman C. Moseley / 85

5.9  Postcard, circa 1940, showing the Charleston Airport / 86

5.10  Badge worn by spectators attending the opening ceremonies for the Cooper River Bridge, August 8–10, 1929 / 86

5.11  Automobiles crossing the Cooper River Bridge during the four-hour toll-free period on opening day / 87

5.12  The Cooper River Bridge's tollhouse, shortly after the bridge was opened / 88

5.13  Tourism guides published in an effort to lure visitors to the city / 90

5.14  The Cooper River ferry terminus in the 1930s / 92

5.15  Photographs from the toll-free celebration, June 29, 1946 / 96

5.16  The Cooper River Bridge as it appeared in the 1940s / 97

6.1  Marion J. Schwartz, director of Charleston County Police, destroys slot machines in Mount Pleasant in circa 1959 crackdown on illegal gambling in North Charleston area / 98

6.2  Photograph taken shortly after the *Nicaragua Victory* collided with the Cooper River Bridge, February 24, 1946 / 100

6.3  Photograph of the *Nicaragua Victory* collision / 100

6.4  A Bailey span used as a temporary "patch" over the damaged sections of the Cooper River Bridge / 101

6.5  Workers installing a Bailey span to patch the section of the Cooper River Bridge knocked out after the collision of the *Nicaragua Victory* / 102

6.6  Postcard showing an aerial view of the Cooper River Bridge (circa late 1950s) looking across at Mount Pleasant / 104

6.7  Summer cottage typical of those found on the Isle of Palms in the early 1900s / 107

6.8  1958 map showing proposed second Cooper River Bridge and Highway 17 bypass / 110

6.9  Program for the groundbreaking ceremony of the second Cooper River Bridge, May 2, 1963 / 112

6.10  The Silas N. Pearman Bridge with the older John P. Grace Memorial Bridge in the background / 113

6.11  Runners and walkers crossing the Silas N. Pearman Bridge as part of the annual Cooper River Bridge Run / 114

6.12  Table: The Cooper River Bridge Run: a Charleston legend / 115

6.13  Dignitaries gathering for groundbreaking of the new multilane Arthur Ravenel Bridge / 121

6.14  The John P. Grace Memorial Bridge listed for sale / 122

# Acknowledgments

Writing the story of the Great Cooper River Bridge has been an extremely long but satisfying undertaking. We thank all of our family and friends for their support of this project. We especially thank Harlan Greene and the staff of the South Carolina Room of the Charleston County Library for all their assistance in researching this book; without their support this project may never have been finalized. We also recognize the countless people whose enthusiasm and encouragement towards this project spurred us on. Thank you—and we hope this story gives you the pleasure it has given us in its telling.

# INTRODUCTION

More than seventy years ago, the skyline of Charleston, South Carolina, was dramatically altered by the construction of the Cooper River Bridge, a tall, spiny, ribbon of a bridge that spans the Charleston harbor. The Cooper River Bridge played a vital role in Charleston's transformation from a poor, deteriorating city into a vibrant and prosperous metropolis. The size and the character of Charleston and its suburbs have changed considerably since the bridge was opened in 1929, and the Cooper River Bridge no longer adequately serves the needs of the growing Charleston area. The structure is old and obsolete, and it has received the dubious honor of being cited as one of the most unsafe bridges in South Carolina. The bridge, along with its larger "sister" structure, the Silas N. Pearman Bridge, erected in 1966, will be replaced. However, over the years, Charlestonians have become accustomed to the Cooper River Bridge. Once it is removed Charleston will lose one of its most significant landmarks.

In 1949, New York architecture critic Elizabeth Mock awarded the Cooper River Bridge the epithet of "the most spectacular bridge in the world." She wrote, "Steep approaches, stupendous height, extremely narrow width, and a sharp curve at the dip conspire to excite and alarm the motorist, even while his changing perspective of the second span gives . . . multiple awareness of the structure that is hurling . . . through space. Perhaps all bridges should be bent at the middle so that no one might traverse them unaware."[1] The bridge's distinctive silhouette appears on souvenirs, postcards, and shirts; its form is used in corporate logos and on official publications. Though it is a relatively new addition to centuries-old Charleston, the Cooper River Bridge has become an icon of the city, much like the Eiffel Tower is to Paris or the Golden Gate Bridge to San Francisco.

The impact of the Cooper River Bridge on Charleston's modern history has been profound. More than three generations of Charlestonians have grown up under the steel rainbow of the Cooper River Bridge, oblivious of a time when crossing the harbor by ferry was an ordeal stymied by storms, fogs, and tides. The bridge funneled hundreds of young families into the booming suburbs east of the Cooper River and played a significant role in shaping that region's growth. Thousands of tourists received their first glimpse of the historic city of Charleston from atop the bridge's lofty spans. The bridge witnessed the arrival of countless vessels at the nearby naval base, and it bid a silent goodbye as the ships and their crews departed the base forever when the shipyard was decommissioned in the mid-1990s. Even today the bridge continues to enrich the wonderful texture of Charleston.

The Cooper River Bridge was opened in 1929, but the complete history of the bridge begins much earlier. Quite unexpectedly, the story of the bridge starts at the beach. The early efforts to finance and build a bridge are wed to the history of the Isle of Palms. By 1900, the Isle of Palms was a popular beach retreat for working and middle-class Charlestonians. The owners of the resort realized that the limited capabilities of the Charleston harbor ferries hampered the island's growth. They envisioned a bridge across Charleston harbor to lure more vacationers to their resort. In 1926 the owners of the Isle of Palms formed the Cooper River Bridge Corporation for the purpose of financing and constructing a cross-harbor bridge. The intent of the company was obvious—to increase the number of visitors to the Isle of Palms via a bridge paid for by toll revenues—yet the bridge promoters also marketed themselves as community-minded businessmen who sought to bolster Charleston's entire economy. After several years of intense political and financial negotiations, construction began on the Cooper River Bridge.

Throughout the planning and construction of the Cooper River Bridge, Charleston businessmen and political leaders heralded the structure as a symbol of the new, modern Charleston. The bridge must be viewed within the context of Charleston in the 1920s, and those forces behind the city's drive towards modernization. For decades after the Civil War, Charleston existed in decay and depression. By the 1920s new industries were pumping much-needed capital into the city, and Charleston's progressive leaders began an orchestrated effort to reform the conservative city. Progressive mayors like John Patrick Grace established the foundations upon which modern Charleston would be built. In this dynamic atmosphere, the Cooper River Bridge was a symbol of progress, a true marvel of modern bridge engineering. City leaders projected onto the bridge all of Charleston's hope for the future. If the great Charleston harbor could be bridged, then certainly Charleston's best years lay ahead.

John Patrick Grace became the driving force behind the development of the Cooper River Bridge, boldly predicting that the structure would spur unparalleled growth and opportunity. As a former mayor and a keen political operative, Grace helped to navigate the Cooper River Bridge project through the intricacies of state and local government. Grace, a johnny-come-lately to the bridge efforts, quickly came to dominate the project. A shameless self-promoter, Grace remarked, "I took the dreams, my own and the dreams of others, and gave them a local habitation and a name 'The Cooper River Bridge.'"[2]

Charlestonians closely followed the construction of the Cooper River Bridge. Local papers provided frequent reports on the highly technical aspects the bridge's design and construction. It was called the most significant engineering event that the city had witnessed since the first passenger steam locomotive, the *Best Friend*, in 1830. Nearly 600 men worked on the project; fourteen workers died in accidents during its construction. For its time the Cooper River Bridge was not the largest or grandest bridge, but its unique curves and steep grades attracted international attention. Upon its completion, engineers called it "the first roller-coaster bridge."

To mark the opening of the Cooper River Bridge, August 8–10, 1929, Charleston hosted a three-day celebration of parades, automobile races, dances, and concerts. The bridge embodied modern Charleston, a point made clear by the local newspapers and civic leaders alike. Newspaper headlines proclaimed, "Charleston Welcomes America!" Grace equated the construction of the bridge to Columbus's discovery of the new world. In the short term the bridge was an economic failure. Several months after the bridge opened, the stock market crashed and along with it the financial aspirations of the Cooper River Bridge Corporation. The bridge failed to deliver profit and failed to bring a great rush of prosperity to Charleston. Throughout the 1930s Charleston, like the rest of the nation, suffered through the Depression, and the Cooper River Bridge became a symbol of unrealized dreams. Charlestonians cursed the bridge's tolls, blaming them for impeding the city's prosperity. Charleston County purchased the bridge in 1941. Soon thereafter, in 1945, the state assumed control of the bridge. When the bridge tolls were removed in 1946 Charleston celebrated, and the city was declared open for business once again.

In the post-toll era, the Cooper River Bridge proved wildly successful. The bridge provided easy access between Charleston and its new bedroom communities east of the Cooper River. Hundreds of families crossed the bridge into the suburbs to begin their pursuits of the American Dream. But, as the population living east of the Cooper River doubled, then tripled, traffic on the bridge became horrendous. By the mid-1950s Charlestonians were again cursing the Cooper River Bridge and demanded a newer, bigger crossing. In 1966, the Silas N. Pearman Bridge was constructed alongside the Grace Bridge. The new bridge offered temporary traffic relief, but Charleston's suburbs east of the Cooper River continued to grow. Today construction is underway on an entirely new bridge across the Cooper River. In place of the two older Cooper River bridges a new, ultra-modern, taller, and grander bridge will stand. The old Cooper River Bridge, once a symbol of modern Charleston, will be destroyed.

Charleston area residents have developed a special relationship with the Cooper River Bridge. Some love it, others hate it; but everyone has a story to tell about the bridge: accounts of Cooper River Bridge footraces, sad stories of accidents or bridge jumpers, and thrilling tales about driving over "the roller-coaster bridge" for the first time. The Cooper River Bridge also has its own story. It is the story of ambitious men and their dreams of profit, and of a city's dreams of prosperity. It is the story of amazing engineering and construction accomplishments, of economic failure and tragedies, and finally of old age and inadequacy. The story of the Cooper River Bridge is the story of Charleston's birth as a modern city. This is the story of the great Cooper River Bridge.

*The Great Cooper River Bridge, Charleston, S.C.* Courtesy of the Charleston County Library

*"Many years ago, I think it was the football season of 1943, the St. Andrews [High School] football team, of which I was a member, was returning from a game with Georgetown. It must have been about midnight when we got to the old Cooper River Bridge. When our school bus got almost to the top of the Town Creek span it ran out of gas and stopped about thirty feet from the top. We were the only vehicle on the bridge, so the team got out of the bus and we all pushed it to the top. Quickly getting back in, we coasted down to Meeting Street, careened around a corner, and pulled into an all-night gas station at the corner of Line and Meeting. Goodness knows what we would have done if the bus had run out of gas in the saddle."*

—Arthur Ravenel Jr., State Senator, July 6, 1998

*The Great*
COOPER RIVER BRIDGE

## CHAPTER ONE

# The Dream

> *I took the dreams, my own and the dreams of others, and gave them*
> *a local habitation and a name "The Cooper River Bridge."*[1]
>
> —John P. Grace

The Charleston peninsula is geographically isolated from the rest of coastal South Carolina. Rivers and swamps create a natural barrier in all directions around the city known, due to the proliferation of church steeples, as the Holy City. Early in its history these hindrances served to protect the city, but in the decades following the War Between the States, the rivers and wetlands only delayed the development of railroads and highways leading to Charleston. By the close of the nineteenth century, bridges and roadways linked the city to thoroughfares inland, but connections to the coastal areas north and south of the city were lacking.

*1.1 Early map of Charleston and vicinity, from* Harper's Magazine. *This map illustrates the numerous waterways that isolated peninsular Charleston from the surrounding mainland. Early in the history of the city, its position served as a strategic advantage. However, with the arrival of railroads, and later automobiles, natural barriers hindered Charleston's efforts to grow and to integrate with the rest of the nation.* Collection of Pamela Gabriel

With the success of Henry Ford's Model T automobile, Americans took to the open road in record numbers. The first "snowbirds" journeyed from the north to Florida in the 1920s and created a land boom in that newly rediscovered tropical state, buying cheap land for winter homes. A great coastal highway was envisioned, and a road network stretching from New England to Florida was initiated. In 1925, old Kings Highway, which eventually became U.S. Highway 17, was part of the Coastal Highway. In South Carolina it swung inland west from Mullins to Florence, around the Santee Swamp along the coast, before entering Charleston from the direction of Columbia.[2] The route completely bypassed the coastal regions north of Charleston. The trip from Charleston to nearby Mount Pleasant, a mere mile by boat or ferry across the Charleston Harbor, was an eighty-mile journey by car over rough dirt roads.

## The First Bridge: The Ashley River Bridge

The Ashley River, the western boundary of the Charleston peninsula, was first bridged in 1808. Tobacco, rice, and lumber produced on James Island and John's Island were transported over the bridge and across the peninsula to the busy wharves of the Charleston harbor for export to distant markets. During the Civil War, Confederate freight trains used the Ashley River crossing to transport troops from the sea

*1.2  The Ashley River Bridge tollhouse as it appeared around 1900. This structure first served as the residence for the operator of a ferry service that crossed the Ashley River. It later housed the keeper of the wooden bridge across the Ashley River. The dwelling burned in 1929.* Photograph from *Charleston Come Hell or High Water* by Robert N. S. Whitelaw and Alice F. Levkoff; courtesy of Alice F. Levkoff

*1.3 Photograph, circa 1890, showing the wooden bridge across the Ashley River. This bridge was built after the hurricane of 1893 destroyed a similar structure built eight years earlier. The bridge was privately owned and operated until the county purchased it in 1921 and was demolished when the steel-and-concrete Ashley River Memorial Bridge was erected in 1926.* Courtesy of the Avery Research Center, College of Charleston

islands into the city. When the Confederate Army was forced to evacuate Charleston, General Pierre G. T. Beauregard ordered the wooden bridge burned in an attempt to hinder the advance of Federal troops. In 1885, a company organized by Beauregard's former chief clerk, John Ficken, built a new wooden toll bridge across the Ashley.[3] An old dwelling located alongside the new bridge, a remnant of an earlier ferry site, served as a tollhouse and residence for the bridge tender. A toll of five cents was levied on pedestrians and ten cents for horse-drawn wagons. Eight years after the bridge's completion, the hurricane of 1893 crashed into the lowcountry, destroying many structures, including the new bridge. Ficken's company promptly rebuilt the bridge. This structure lasted another thirty years, later accommodating automobiles for a toll of fifteen cents.

In 1923 construction began on a new, toll-free, concrete bridge over the Ashley River after plans for the structure were formulated during the administration of Charleston mayor John P. Grace. Federal funding assisted in financing the bridge, which was designed by James L. Parker, the engineer of the first bridge over the Santee River at the far northern end of Charleston County. The new Memorial Bridge formally opened on May 5, 1926, during the administration of John Grace's political nemesis, Mayor Thomas Stoney. Ficken's old wooden bridge, which had its toll removed when the county bought the structure in 1921, was torn down shortly after the Memorial Bridge was completed. The Memorial Bridge, built at a cost of $1,250,000, opened amid great fanfare and celebration and was dedicated to those who served during the Great War (World War I). Charleston's newspaper, the *News and Courier*, published a sixty-eight page addition extolling, in great detail, the marvels of the "widest and handsomest bridge in the South."[4] The bridge is still in use. The 34-foot roadbed originally featured four lanes for traffic and sidewalks four and a half feet wide (today the bridge has only three lanes leading out of the city). The main span has mechanically operated draws providing 110-foot horizontal clearance for passing boats. Including the approaches, the Memorial Bridge is almost one mile long. A second bridge over the Ashley River, the T. Allen Legare Jr. Bridge, was constructed alongside the Ashley River Memorial Bridge in 1960.

After the Ashley River Memorial Bridge opened, new suburbs sprouted up west of the Ashley River (simply known as West Ashley). Here Charlestonians built modern homes in new neighborhoods such as The Crescent and Wappoo Heights. Though the roads between Charleston and Savannah remained bumpy and unpaved, the Ashley River Bridge helped to fulfill part of the dream of a north-south route linking Quebec to the Florida Keys.[5] The Santee River Bridge had been completed several years prior to the opening of the Ashley River Bridge, and the two bridges were important links in the Coastal Highway. With the Ashley River bridged, attention was focused on the eastern edge of the Charleston peninsula; the Cooper River was still without a bridge.

## The Coney Island of the South

Today the area east of the Cooper River, or East Cooper, is home to Mount Pleasant, a burgeoning suburb and one of South Carolina's largest cities. But in the early 1900s the town was a sleepy village whose residents were shipbuilders, shrimpers, and farmers. At the time, other significant communities in East Cooper included unincorporated Christ Church Parish and tiny McClellanville, north of Mount Pleasant. Further north of McClellanville was Georgetown, sixty miles from Charleston. A trip to Charleston from McClellanville via the Cooper River ferries was an day-long journey. Though many East Cooper residents looked forward to the convenience of a bridge, the undertaking was a massive, expensive project. By itself, East Cooper's small and dispersed population did not offer the economic justification for such a project. Rather, the early dreams of bridging the Cooper River were linked to the ambitious plans of the owners of the Isle of Palms beach resort.

While Sullivan's Island successfully defended the Charleston Harbor and served as a summer retreat for affluent Charlestonians, its northern neighbor, the Isle of Palms,

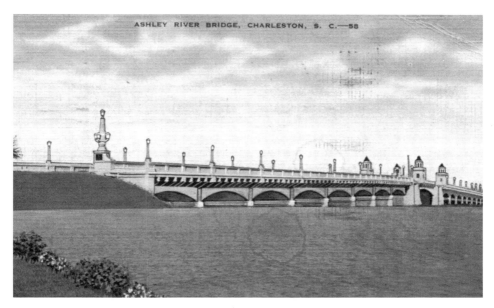

*1.4  Postcard, circa 1930, depicting the Ashley River Memorial Bridge. When it was completed in 1926 this bridge, dedicated to the memory of those who fought in World War I, was described as "the widest and handsomest bridge in the south." In 1993 new lights were installed to resemble the city's old gas burning street lamps.* Collection of Pamela Gabriel

*1.5  The Santee River Bridge, located between McClellanville and Georgetown, S.C., was completed in the mid-1920s. This bridge and the Ashley River Memorial Bridge, were important links in the Coastal Highway.* Courtesy of the Charleston County Library

remained an isolated tropical wilderness. The uninhabited barrier island was privately owned and secluded until late in the nineteenth century. The seven-mile-long island was originally called Hunting Island due to the abundance of wildlife, but for most of its history the island was known as Long Island. Early in its history, the Sewee Indians inhabited the island. Thomas Holton, a planter from Barbados, received a grant to the land from the Lords Proprietors in 1696, and upon his death the grant passed to his son. Reportedly Long Island served as a haven for pirates whose buried treasures have yet to be uncovered among the sand dunes. During the Revolutionary War, British troops encamped on the island attempted to storm patriot soldiers defending Charleston Harbor from new fortifications on nearby Sullivan's Island. The treacherous currents of Breach Inlet, which separates the two islands, thwarted their attack. Long Island remained largely undeveloped until 1898, when Dr. Joseph S. Lawrence, who introduced the electric trolley to Charleston, purchased the island with a plan to construct a resort along a sandy stretch of beachfront. His first order of business was renaming the island to the more exotic Isle of Palms. Lawrence's visionary ideas forever altered the quiet of this island paradise.

Late in the nineteenth century a growing middle class of shopkeepers and clerks emerged in Charleston. Oppressive summer temperatures made this group anxious for an escape from the heat of the city, a privilege previously available only to the rich. Lawrence, quick to recognize this need, set out to develop the Isle of Palms into an affordable beach resort similar to other middle-class beaches such as Coney Island in New York and Atlantic City in New Jersey. Natural obstacles stood in the way of Lawrence's ambitious scheme: namely the Cooper River, Cove Inlet (between Sullivan's Island and the Mount Pleasant mainland), and Breach Inlet. The first challenge was to devise a means for transporting visitors to the island resort. Naming himself as president, Lawrence established the Charleston and Seashore Railroad. He also purchased an existing ferry service and built railway bridges as part of his grand scheme to integrate ferry and rail service to transport vacationers to the Isle of Palms.

Ferries had been a presence in Charleston Harbor and the Cooper River since 1748, the year in which Henry Gray was granted a charter to operate a ferry service between Charleston and Christ Church Parish. In 1765, a second charter was granted to Clement Lempriere to run a ferry between Charleston and Hobcaw Point, near Remley's Point. Five years later Andrew Hibben received a charter to operate a ferry between Charleston and the village of Mount Pleasant; his descendants operated this service until 1850. Other ferry services, such as Milton's ferry, Hunt's ferry, and Mathewes' ferry, operated from the area of Hog Island. Many of the old ferry services have left their mark in Mount Pleasant street names, such as Hibben (Ferry) Street and Mathis (a corruption of Mathewes) Ferry Road. In 1856, three men (Charles Jugnot, C. D. Carr, and Oliver Hilliard) purchased Barksdale Point, near the old village

*1.6 Cove Inlet Bridge, between Mount Pleasant and Sullivan's Island, circa 1920. This bridge across Cove Inlet served the electric trolleys that ran between the ferry slips in Mount Pleasant and the Isle of Palms. The Cooper River Ferry Commission replaced it with an automobile crossing in 1924. The newer Cove Inlet Bridge, known as the Pitt Street Bridge, was replaced by the Ben Sawyer Bridge in 1947.* Courtesy of SCANA

*1.7 The trolley station on the Isle of Palms, circa 1910. Visitors to the Isle of Palms beach resort disembarked from the trolley at this small station near the resort's picnic grounds.* Collection of Pamela Gabriel

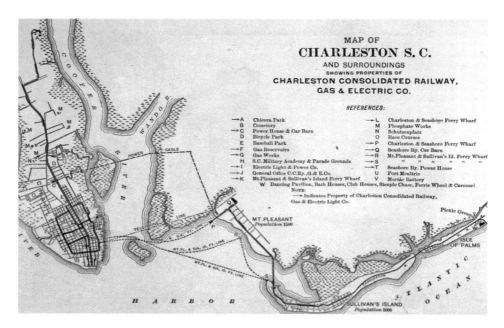

*1.8 Ferry map from early 1900s published by the Consolidated Railroad, Gas and Electric Company. The map shows the routes of the ferries and trolleys from Charleston to the Isle of Palms. The Consolidated firm, owned by Dr. Joseph S. Lawrence, operated the ferries and trolleys.* Courtesy of the Charleston County Library

of Mount Pleasant, from William McCants and together formed McCants Ferry Company. The new ferry service began making two round trips daily to Charleston.[6] The company later changed its name to Mount Pleasant and Sullivan's Island Ferry Service, which Joseph Lawrence purchased in 1898.

Visitors to Lawrence's Isle of Palms resort embarked upon their journey from the Charleston wharves and crossed the harbor to Mount Pleasant via ferries. Here they were deposited at the end of Ferry Street, where they boarded the electric trolley cars of the Seashore Railroad and traveled through the village of Mount Pleasant on the newly laid tracks on Pitt Street. At the end of Pitt Street, the trolley cars continued across mile-wide Cove Inlet and onto Sullivan's Island via a new low-level trestle bridge, which was equipped with a movable span that swung open for marine traffic. This bridge replaced an earlier footbridge that was situated there prior to the Revolutionary War.

The trolley reached Sullivan's Island near Fort Moultrie, where the powerhouse for the electric trolley was located. The trolley cars traversed Sullivan's Island along Middle Street before turning slightly and proceeding to Breach Inlet along Railroad Avenue (now called Jasper Boulevard). The cross streets of Sullivan's Island still carry the numbered trolley stops along the route—Stations 18, 19, 20, and so forth.[7]

1.9 An electric trolley used by Consolidated Railroad, Gas and Electric Company to transport vacationers to the Isle of Palms during the early 1900s. Trolley service carried vacationers from the ferry slips in Mount Pleasant, across Sullivan's Island, and ended at the Isle of Palms beach resort. Courtesy of SCANA

The Hotel Seashore, Isle of Palms,
Charleston, S. C.

*1.10 Postcard from the early 1900s showing the Hotel Seashore on the Isle of Palms. Originally the porches around the hotel were glassed in, but after strong winds destroyed the windows in a storm, the porches were left open for guests to sit on rockers and enjoy the beach breezes.* Collection of Pamela Gabriel

Isle of Palms, Charleston, S. C.

*1.11 The Ferris wheel at the Isle of Palms, shown in this postcard from 1913. The 186-foot-high wheel, built to resemble the one showcased at the 1893 World's Fair in Chicago, offered a spectacular view of the Charleston harbor and surrounding areas. High winds destroyed the Ferris wheel in 1946.* Collection of Pamela Gabriel

*1.12 The* Commodore, *a double-ended ferry and the largest ship in the ferry fleet, was capable of carrying up to 1,200 passengers during its fifteen-minute crossing of the Cooper River. In the 1920s it was equipped to transport vehicles. This image appeared in an advertising brochure for the Isle of Palms, circa 1905.* Courtesy of the Charleston County Library

Another trestle bridge carried the trolley over Breach Inlet between Sullivan's Island and the Isle of Palms and deposited passengers a short walk away from the beach for a day of sunbathing and swimming. The entire journey from the sweltering city of Charleston to the cool breezes of "Parm" Island took approximately one hour—and the trip was half the fun.[8] Hats blew off along the route, and the rocking, bouncing, jarring last car of the trolley was popular with the young children who called it the "crack-the whip" ride. The cost for the round trip excursion was twenty-five cents.

By August 1898, a regular ferry schedule was established with nine round trips between Charleston and Mount Pleasant made daily and seven on Sundays. A year later Lawrence merged the Charleston and Seashore Railroad with the City Railway company, which operated Charleston's streetcars, and formed the Consolidated Railroad, Gas and Electric Company, simply known as Consolidated. The popularity of the outings escalated, and vacationers from as far away as Columbia and Augusta took advantage of low railroad rates to spend a day at the Isle of Palms. The fifty-room Hotel Seashore was built to accommodate overnight travelers. A restaurant, a dance pavilion, a music hall, a zoo, and amusements were added, including a 186-foot Ferris wheel resembling the one invented for the 1893 Chicago World's Fair. Other attractions included a carousel and a steeplechase ride. In 1905, Consolidated began leveling the sand dunes and building summer cottages with plans for year-round residences similar to those on neighboring Sullivan's Island, which had been home to fashionable vacation villas since the 1850s but had attracted vacationers since the early 1800s.

*1.13 Cover of early 1900s promotional brochure, published by the Charleston Consolidated Railway Company, for the beach resort on the Isle of Palms.* Courtesy of the Charleston County Library

## Sullivan's Island

Sullivan's Island is located to the northeast of the Charleston peninsula. The southern tip of the boomerang-shaped island extends into the Charleston Harbor, and the northern section stretches into the Atlantic Ocean. This vantage point, overlooking the harbor and the ocean, created an excellent strategic defensive site, and Sullivan's Island served in this sentry capacity for nearly three hundred years.

In 1674, Captain Florence O'Sullivan, a member of the Provincial Parliament of 1672, was outfitted with a signal cannon and deposited on the island as a lookout for Spanish warships. It was said that O'Sullivan was a logical choice, given his unpleasant disposition.[9] Over time the island became identified with Captain O'Sullivan and eventually came to bear his name. One hundred years later Fort Sullivan was built on the site of Captain Sullivan's lookout. Known as Palmetto Fort because of its palmetto log construction, the fort was still incomplete early in the Revolutionary War when the British attempted to sail into Charleston Harbor and seize control of the city. The British navy attacked the patriots defending the harbor from the unfinished fortification. Fortunately, the palmetto logs and sand of the new fortification absorbed the British cannon fire, which inflicted surprisingly little damage to the fort. The hero of the battle, in which the British suffered one of its major defeats of the war, was Colonel

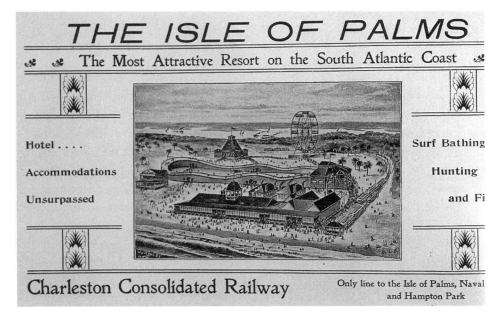

*1.14 Advertisement for the Isle of Palms that appeared in a 1903 tourism brochure titled "Progressive Charleston." The illustration shows the resort's top attractions: the Ferris wheel, pavilion, hotel, and undulating steeplechase ride.* Courtesy of the Charleston County Library

William Moultrie. The rebuilt fort was renamed in his honor, and grateful South Carolinians would eventually incorporate the palmetto tree into the state flag. The fort fell victim to a hurricane in 1804 but was rebuilt five years later. Fort Moultrie continued to serve as an integral part of the American defense system until it was decommissioned in 1947.

In addition to playing a notable role in U.S. military history, Sullivan's Island was a disembarking point for many African slaves. Thousands of sick Africans were deposited on the beaches of Sullivan's Island before being sold into slavery. As early as 1707, the island was designated as a quarantine area, or "lazaretto," for captured Africans being brought into Charleston. The "pest house" located near the cove of the southern point of the island served as a holding area to ensure that contagious diseases were not brought into the city by slaves.[10]

At the height of Charleston's antebellum prosperity, wealthy residents escaped the miasma of the city by enjoying the sea breezes on Sullivan's Island. After Fort Moultrie became an army post in 1809, a viable community developed around the fort; a town was incorporated in 1817. Architect Robert Mills described the place, already functioning as a summer resort, in 1826: "The village is called Moultrieville, in honor of the defenders of the fort . . . 200 houses, all of wood . . . are occupied . . . during the summer."[11] St. Paul's Episcopal Church was founded on the island in 1819, and a

Presbyterian church in 1826. One of the most famous of those who served at Fort Moultrie was writer Edgar Allan Poe, who later described the island in his story "The Gold Bug":

> The island is a very singular one. It . . . is about three miles long. Its breadth at no point exceeds a quarter of a mile. It is separated from the mainland by a scarcely perceptible creek, oozing its way through a wilderness of reeds and slime, a favorite resort of the marsh hen. The vegetation, as might be supposed, is scant or at best dwarfish. No trees of any magnitude are to be seen. Near the western extremity, where Fort Moultrie stands, and where are some miserable plain buildings, tenanted during the summer by the fugitives from the Charleston dust and fever, may be found the bristling palmetto: but the whole island, with the exception of the western point, and a line of hard, white beach on the seacoast, is covered with a dense undergrowth of sweet myrtle.

The island recalls Poe's military stay in the street names of Poe Avenue, Gold Bug Avenue, and Raven Drive. Other street names offer insight into the island's history. I'On, one of the island's main thoroughfares, is named for politician and War of 1812 veteran Colonel Jacob Bond 'Ion (pronounced *EYE-on*; at a later time the accent shifted to after the "I"). By all accounts Colonel 'Ion was a very large and jolly man. He was affectionately called "Old Uncle" and his home was a favorite among young soldiers stationed at Fort Moultrie.

By the mid-1800s the island witnessed a seasonal influx of affluent Charleston citizens anxious to enjoy cooling ocean breezes.[12] Families packed furnishings, clothing, linens, silverware, china, livestock, pets, horses and carriages, and food and wine, and, along with their household servants, made preparations for "summering" on the island. Barrels, trunks, and boxes were loaded on wagons, carted to the wharves on the Cooper River, stacked on "rum boats," and transported to the island. Three months later, family and servants would close up their cottage and repack trunks, barrels, and boxes and return to the city.

Most of the early cottages were destroyed during the War Between the States. After the war Charlestonians were permitted to return to the island to rebuild their homes. Some of the new homes were quite grand, such as the mansion of John Henry Devereaux, a Charleston architect, who in 1869 designed Stella Maris Roman Catholic Church, located adjacent to Fort Moultrie. After years of redevelopment, cozy cottages dotted the island in both Moultrieville and the new section, Atlanticville. The Atlanticville community, located in the middle of the island, sprang up around the New Brighton Hotel, built in 1881. This elegant, 112-room beachfront hotel was built by wealthy northerner John Chisholm and was located around what is today Station 22½.

FORT MOULTRIE.—CHARLESTON IN THE DISTANCE

1.15 *The village of Moultrieville on Sullivan's Island as it appeared circa 1850. The village developed around Fort Moultrie, a U.S. army fort, and was home to two churches and the summer cottages of many wealthy Charlestonians.* Collection of Pamela Gabriel

All houses built on Sullivan's Island were built on land leased from the state.[13] Lots were leased for a period of fifty or seventy-five years, after which a change in title was necessary. Many leaseholders transferred the property to a family member as a legal gesture, thus ensuring that the property would remain in the family. The state also limited the number of lots for lease to 500, effectively capping growth of the island.[14] For these reasons Sullivan's Island remained a quiet, family beach resort. The privately owned Isle of Palms, by contrast, offered unlimited opportunity for growth.

## New Owners, New Dreams

After successfully developing and marketing the Isle of Palms as a beach resort, ill health forced Joseph Lawrence to retire. In 1913, his Consolidated firm was sold to a new company, the Charleston Isle of Palms Traction Company. The new president was James Sottile, whose Sicilian family built the first cottage on the island in 1897. The Sottile family had acquired significant wealth in the amusement industry; it owned and managed a number of theaters in Charleston. The new company continued to market the seaside resort, promoting special events and attractions such as motorcycle races, to lure crowds to the island.

During the summer months, vacationers swarmed onto the ferryboats bound for the beach. James Sottile realized that a faster, more reliable means of transportation

*1.16  The Atlantic Beach Hotel on Sullivan's Island, shown here in the early 1900s, was built in 1881 as the New Brighton Hotel and later changed its name. It was known for its grand summer parties. The hotel burned in 1925 with a heat so intense that some of the surrounding beach sand fused into glass particles.* Courtesy of SCANA

*1.17  Bathers on Sullivan's Island. The hotel in the background is most likely the Atlantic Beach Hotel.*
Photograph from *Charleston Come Hell or High Water* by Robert N. S. Whitelaw and Alice F. Levkoff;
courtesy of Alice F. Levkoff

was needed to bring visitors to his resort. In 1913, he put forth the first proposal for bridging Charleston Harbor and sought an exclusive franchise, or monopoly, for such a bridge. In 1913, Sottile asked South Carolina's U.S. Senator Benjamin Tillman to introduce legislation in Congress that would grant him a franchise for his bridge. The action irked many Charlestonians, who felt that Sottile was bypassing local authorities in his quest to build a bridge. Though his plans were vague, Sottile proposed a low-level concrete bridge for carrying trains across the harbor and another bridge across Shem Creek. To accommodate shipping traffic, Sottile's design called for several lift spans, sections mechanically raised and lowered in a horizontal position. Fearing that this structure would interfere with the harbor, the Charleston shipping industry immediately voiced opposition, as did the U.S. Navy, which had recently opened a shipyard along the Cooper River. Critics argued that Sottile's bridge plan was too vague and hastily proposed, and that the mechanical lift spans it employed were prone to failure. Ironically, one of those responsible for thwarting Sottile's project was Charleston Mayor John Grace, the man who would eventually assume the lion's share of praise for bridging the Cooper River sixteen years later.

Vowing to "build the bridge if it took his last dollar," James Sottile's existing financial troubles were intensified by the brief depression that followed World War I.[15] His bridge plan lacked local support and died a quick and silent death. In 1923 Sottile was sued by a ferry passenger, and the resulting personal injury judgment of $10,000

*1.18 The* Nansemond, *one of several ferries that transported automobiles across the Cooper River during the 1920s. During the 1920s East Cooper trolley service was phased out, made obsolete by automobiles.* Courtesy of Greyscale Fine Photography Center, Charleston, S.C.

bankrupted him. Sottile's ferry service across the Cooper River was halted in February 1924 when the county sheriff seized the ferries by court order. At the time there was no other ferry service in operation to replace Sottile's failed business. For the first time in over 150 years Charleston was without a cross-harbor ferry. A government-sponsored ferry service, the Cooper River Ferry Commission, was created, and cross-harbor ferry service for both passengers and automobiles resumed in March 1924.

When the Cooper River Ferry Commission inherited the responsibility of providing cross-harbor ferry service it found that the existing infrastructure was in a terrible state of disrepair.[16] Ferry slips and wharves were decrepit. The East Cooper trolley service had been halted. The Cooper River Ferry Commission immediately began improving the infrastructure. It repaired the trolley facilities, and trains were put into limited service in 1925. The commission also began building new automobile roads and bridges as part of a rapid phase-out of the trolley system. All trolley service was discontinued by 1927. The commission refurbished the dilapidated ferry slips in Charleston and built new ferry facilities on Hog Island in Mount Pleasant (near present day Patriot's Point). To provide better access to the ferry slips at Hog Island, the Cooper River Ferry Commission successfully lobbied the state to construct the first automobile bridge over Shem Creek in Mount Pleasant.[17] The commission also lobbied the War Department to help finance a new automobile bridge across Cove Inlet, between Mount Pleasant and Sullivan's Island, arguing that the bridge would be used to serve the military community at Fort Moultrie. Though the War Department rebuffed its efforts, the Cooper River Ferry Commission did succeed in securing state and federal financing for the project. The new concrete-and-steel Cove Inlet Bridge, an automobile crossing, was begun in 1924. The state highway department assumed control of the completed bridge in 1927.[18] This bridge was closed in 1945 when the new bridge to Sullivan's Island, the Ben Sawyer Bridge, was opened.

The Cooper River Ferry Commission was a self-financing governmental agency. In the legislation that created the commission, toll rates were locked at five cents per person. During the first full year of operation the commission lost nearly $20,000. In 1926, the state agreed to raise the rates to fifteen cents per person and fifty cents per automobile, including driver. The increased toll made the ferry operations profitable. For the next several years the Cooper River Ferry Commission dramatically increased its passenger volume, despite the fact that the Isle of Palms remained in a state of neglect following Sottile's bankruptcy. In 1925 the ferries transported about 380,000 passengers and 32,000 automobiles; by 1927 the annual passenger volume on the ferries had increased to 526,000 passengers and 100,000 automobiles. In 1928 ferry rates

*1.19 Automobiles disembark from one of the Cooper River ferries. Though beach vacationers supplied the bulk of revenues for the ferries, East Cooper citizens depended on the service for their frequent commutes into Charleston. In her book* Through a Turnstile into Yesterday, *Mary Sparkman remembers the "harmonious relations between the daily regular passengers and the pilots of the ferry-boats." If a commuter was running late, it was not uncommon for the ferry pilot to delay departure until they arrived. The ferries also offered music on Sunday and "hop-nights" with live music to entertain the passengers.* Courtesy of Greyscale Fine Photography Center, Charleston, S.C.

*1.20 The hotel and pavilion on the Isle of Palms in the 1920s. Throughout the late 1920s the island remained an immensely popular beach vacation spot for both Charlestonians and out-of-town visitors.* Courtesy of Sandy Knisley, from the collection of Robert Legaré Coleman

were lowered to ten cents per person, but the increase in the number of passengers contributed to a net gain in profits.

When the Isle of Palms was sold to a group of investors in 1925, the resort was in shambles. Cash-strapped Sottile had put little money into the facilities. While the Cooper River Ferry Commission shouldered the responsibility for improving the ferry slips and providing new transportation infrastructure, the new resort owners improved facilities and developed a new vision for the Isle of Palms. Their prospectus included maps and outlined plans for hotels, restaurants, lakes, and golf courses. All of the group's planning centered on the completion of a privately owned and operated bridge across Charleston Harbor that would allow unhindered access to their resort.

The decision to pursue a private bridge was motivated by several factors. Ferry boats remained the only means of crossing the Cooper River. Ferries such as the *Lawrence*, the *Palmetto*, and the *Sappho* shared the busy harbor with merchant ships, fishing boats, and naval vessels and were often delayed by shipping traffic, naval maneuvers, tide changes, and the weather. Charleston also suffered when Sottile's ferry service was halted in 1924, demonstrating to many the city's heavy reliance on the ferry service. Furthermore, trolley service in East Cooper had ceased to operate by 1927. Now automobiles, not trolleys, were the only means of traveling from Mount Pleasant to the Isle of Palms. The owners of the Isle of Palms were acutely aware of the increasing financial success of the ferries. Each year since its inception the Cooper River Ferry Commission had realized an increase in the number of passengers and automobiles it transported. Bridge promoters believed that a toll bridge across Charleston Harbor would not only provide a reliable means of transporting tourists to the Isle of Palms but would also be profitable. The developers dreamed that the remote island would be a grand beach resort. All that was needed was a bridge to bring in the throngs of waiting vacationers.

## CHAPTER TWO

# CHANGE COMES TO CHARLESTON

The early 1900s were a time of great change both for the nation and for Charleston. The years between 1865 and 1920 were hard times for Charleston. The Civil War left the city physically devastated and financially shattered. Even before the war, however, Charleston was no longer the significant shipping center that had once rivaled the ports of New York, Philadelphia, and Boston. Following the war Charleston experimented with a handful of industrial projects designed to boost the city's fortunes. Most of these plans met with minimal success or outright failure. Railroads had offered a brief flicker of hope by bringing goods to the harbor, but they never amounted to a true boon for the city, largely due to their restrictively high rates. The South Carolina Interstate and West Indian Exposition, which opened in 1901 on the site of the old Washington Race Course (now the site of the Citadel and Hampton Park), hoped to attract worldwide attention to Charleston and reestablish trade with the Caribbean but proved a financial failure. Poor legislation effectively wiped out Charleston's nascent phosphate industry. The boll weevil and economic malaise dealt a crippling blow to the South's cotton industry, reducing the flow of what little cotton still passed through Charleston's port facilities. Even nature seemed determined to punish the deteriorating city with a series of disasters—a destructive cyclone in 1885, a deadly earthquake in 1886, and the Great Hurricane of 1893. Another deadly hurricane assaulted the lowcountry in 1911, destroying the few remaining rice plantations.

For most of the late nineteenth and early twentieth centuries, Charleston remained a poor, decayed, and shabby relic still bearing the scars from a devastating war of decades earlier. Sanitation conditions in the city were deplorable and made worse by the low-lying terrain that created drainage problems. Malaria outbreaks remained a threat until the late 1920s, when city health inspectors began spraying gasoline into ditches and swampy areas in an attempt to control the mosquito population. The city's filth was exacerbated by the common practice of keeping livestock within the city limits. Residents drank from shallow wells that were often contaminated by numerous outdoor privies. Not until the late 1910s did Charleston have a clean water supply, and open sewers ran in parts of the city until the early 1930s. In 1918 the U.S. Public Health Service estimated that there were more than 70,000 rats in Charleston, roughly one for every city resident. The infant mortality rate for Charleston in 1920

*2.1 An impoverished area in Charleston in the early 1900s. Charleston in the early 1900s was poor, decayed, and dirty. Livestock were kept within city limits, and some streets had open sewers.* Courtesy of the Avery Research Center, College of Charleston

was horrendous: 126 deaths per 1,000 births, compared to a nationwide average of 86 deaths per 1,000 births. The gap between Charleston's blacks and whites was even more dramatic: 197 infant deaths per 1,000 births for blacks, compared to 45 deaths per 1,000 for whites.[1] Until the 1920s, city roads were poorly paved, and many streets were not paved at all. During fierce storms and floods, streets could become rivers of mud and trash. Buildings and residences were decaying, yet Charleston remained "too poor to paint but too proud to whitewash." Poetic Charlestonians preferred to see the layers of dirt and ruin as a soft patina upon the city, or to see their city as artist and writer Elizabeth O'Neill Verner described it, merely "mellowed by time."[2] Conservative Charlestonians clung to tradition and fought fiercely with progressive leaders who sought to modernize and improve the city. Between 1910 and 1920 a series of violent city elections pitted conservatives against progressives and demonstrated the great resistance of many Charlestonians towards change. However, beginning in the early 1900s the foundations of modern Charleston were established.

## The Naval Base and Tourism

In a spirit of forgiveness for the sin of secession, the federal government commissioned a naval shipyard in Charleston in 1901. A year later President Theodore

Roosevelt toured the new site. In 1909, President William Howard Taft also favored Charleston with a visit. The shipyard gave the city a much-needed economic boost, creating many new jobs. The construction costs alone brought an estimated $15 million into the city. The annual payroll for the facility was approximately $500,000 ($9.3 million in year-2000 dollars), and by 1913 the naval shipyard provided 20 percent of all wages earned in Charleston.[3] The construction of the naval yard in Charleston was an immensely significant event, not only from an economic standpoint but also from a political perspective. With an influx of federal funds, Charleston was symbolically reincorporated into the nation proper. After the naval base construction, the Charleston *News and Courier* proclaimed "the resumption of diplomatic relations between the White House and Charleston."[4]

In 1912, federal funds were once again directed to Charleston when the government decided to build an immigration station in the Charleston harbor to relieve the crowded conditions at Ellis Island in New York Harbor. The building never processed a single shipload of immigrants. The failure of the immigration station can be attributed to the lack of large-scale industrial employers in the South at a time when manufacturing thrived in the Northeast and Midwest. The South remained a largely agrarian society, and black southerners already held most of the limited number of low-paying positions that traditionally attracted immigrant workers. In fact, so bleak were employment opportunities that during this period, thousands of poor southern blacks migrated north in pursuit of better-paying jobs and to escape the oppressive Jim Crow laws. The net result was a black exodus from the South. By 1920 Charleston's white population exceeded its black population for the first time in more than a century.

While the naval shipyard did much to help the city's sagging economy, Charlestonians continued to seek creative means to generate revenue. Establishments promoting gambling, prostitution, and drinking flourished, thanks to an influx of shipyard workers and ever-present seamen. Conditions in the city were such that one observer compared the city to a "wild west" town. Illegal bars known as "blind tigers" were openly acknowledged, and the fines levied upon these businesses helped fill the city coffers. Even respectable citizens carried firearms. While most city officials preferred to ignore these infractions, the rest of the state was not as tolerant. In 1915, as a concession to newly elected governor Richard Manning, Charleston mayor John Grace initiated a program of liquor law enforcement. Included in the agreement was "the closing of the blind tigers on Sunday and at mid-night, shutting down the slot machines, eliminating lighted signs advertising brothels, and attempting to restrict the sale of liquor to minors and those already intoxicated."[5]

The 1920s were a time of rapid change for America. Mass media communication brought about cultural changes, making life less regional and more diverse. The first

*2.2 The old immigration station, Charleston, S.C. Not long after it was constructed in 1912, the station was converted to a clothing factory, which closed in 1923. The structure served as the Charleston County jail from the 1930s until 1968 and was jokingly referred to as the Seabreeze Hotel. The structure remained empty and neglected until a marine contracting company purchased the property in the 1980s. Seabreeze Development purchased and restored the building in the late 1990s, as shown in this contemporary photograph.* Courtesy of Seabreeze Development

commercial radio broadcast was transmitted in 1920, and by 1930 one out of four American households owned a radio. Charleston's first radio station went on the air in 1930. Nationwide, people from vastly different backgrounds were laughing to the antics of *Amos and Andy*, snapping their fingers to the sounds of New Orleans jazz, or dancing the Charleston.[6]

Along with the rest of the nation, Charlestonians clamored for modern conveniences: automobiles, electrical appliances, radios, and the latest fashions. Through the marvels of mass marketing and nationwide distribution, individuals sought a higher standard of living as depicted in magazine advertisements and movies. Thanks to the perfection of installment buying, these material items were within reach of the ordinary consumer. Many new conveniences promised a savings of time and labor and a better quality of life. As a result, Americans spent more time on amusement and recreation. In 1920, the average American work week was 47.2 hours long; by 1930, the time spent in the workplace decreased considerably. Quite literally, the nation began to arrive at Charleston's door. With increased leisure time and the mobility afforded by personal automobiles, Americans could travel. It was not long before Charleston leaders came to realize the value of tourism.

For centuries, Charleston's geography separated it from the rest of the state. Over time Charlestonians had developed social and political attitudes that also separated it from other regions of South Carolina. The profound social and cultural changes of the 1920s narrowed the gap that separated Charleston not only from the rest of the state but also the rest of America. Motion pictures, telephones, national magazines, radio, and tourism helped awaken Charleston. The city's younger leaders began to advocate modernization—a revival that had started in the early 1890s. Charleston began dramatic efforts to improve sanitation and build up its infrastructure. Its leaders began aggressively courting business and tourists. Slowly, and often begrudgingly, Charleston was turning from its stubborn indifference and detachment from the modern world towards more progressive ideas.

As city leaders attempted to find ways to boost Charleston's economy, the answer arrived in the form of the first wave of tourists to descend upon the city. Visitors discovered "America's best-kept secret"—a charming but shabby, antebellum city. Tourism in Charleston and other southern cities was boosted by a popular literary movement that came to be known as the Southern Renaissance, a term applied to the works of those artists, poets, and writers who were raised in the South and whose work reflected southern culture and lifestyles. One such writer, Julia Peterkin, a South Carolinian, received the Pulitzer Prize in 1929 for her novel *Scarlet Sister Mary*. The nation devoured the works of southern writers such as Thomas Wolfe and William Faulkner, whose works emerged in the late 1920s and 1930s. Northern fascination with "things Southern" and the affordability and availability of the automobile attracted visitors to the South in unprecedented numbers.

The lasting effects of tourism altered the city more than the Union bombardment or the gale forces of hurricane winds ever had. City leaders quickly recognized the value of these visitors to the city, and, while Charlestonians may not have liked "Yankees," they certainly welcomed Yankee dollars. By the late 1920s tourism had become Charleston's biggest industry, and each year an estimated 47,000 tourists were spending $4 million dollars in the city. Despite the claim by the *News and Courier*, and later repeated by Mayor Thomas Stoney in 1929, that "unlike Aiken and Camden, Charleston is not a 'tourist town,'" the industry had an immediate impact upon a city that had previously demonstrated a long tradition of resisting dramatic change.[7]

Charleston was the last significant American city to adopt an electric trolley system, doing so in 1897. The city's first "skyscraper," the eight-story Peoples Building, was not erected until 1911, at a time when New York City was building the sixty-story Woolworth Building. Tourism brought change, and the city was particularly quick to accommodate the new visitors and their money. Mayor John Grace's administration donated land for the Fort Sumter Hotel, built in 1923, overlooking the harbor and White Point Gardens. The city actively encouraged the construction of the

*2.3 Postcard, ca. 1920, showing the Peoples Building on Broad Street (the tall structure at center right). When the People's Building, Charleston's first "skyscraper" (an eight-story bank) was built in 1911, some residents heralded the city's move into the twentieth century. Others saw the building as an sign that Charleston was moving away from tradition. The elaborate overhanging cornices shown in this image were destroyed by a tornado in 1938. In 2001 the building underwent massive renovations.* Collection of Pamela Gabriel

*2.4 Postcard from the early 1900s showing the Fort Sumter Hotel. In 1923, when the city reclaimed marshland near White Point Gardens and created Murray Boulevard, a parcel of the new land was donated for a new hotel.* Collection of Pamela Gabriel

*2.5 The Charleston Hotel as it appeared in the early 1920s. Built in 1836 and located at Meeting and Pinckney Streets, it was one of Charleston's grandest hotels. Its destruction in 1960 was considered a great loss for the city. A bank now occupies the site.* Collection of Pamela Gabriel

thirteen-story, three-hundred room Francis Marion Hotel, the city's tallest structure, opposite Marion Square Park in 1924. The building remains the tallest structure on the Charleston peninsula.[8]

Two-hundred-year-old streets were widened and paved to accommodate automobile traffic. Service stations replaced stately dwellings. Tourists carted away finely carved paneling, elegant fireplaces, elaborate ironwork, and other architectural treasures as souvenirs. The active interest of outsiders caused Charlestonians to take a new look at their aging, deteriorating city, and citizens began to appreciate more fully the city's historical, artistic, and architectural treasures. Native son Dubose Heyward, author of *Porgy*, which was adapted into the folk opera *Porgy and Bess*, looked anew at his home town through an artist's eyes and remarked, "They tell me she is beautiful, my city."[9] Concerned Charleston women seeking to limit the plundering and destruction of grand old homes created the Society for the Preservation of Old Dwellings. The organization led the way in an emerging national trend for preservation of historic structures. Through the efforts of caring citizens, Charleston retained many of its charming old homes, gardens, streets, and neighborhoods.

The prosperity of the 1920s and the increase in visitors to Charleston spurred the owners of Isle of Palms to step up their efforts toward rebuilding the deteriorating resort and to set in motion the processes necessary for building a harbor bridge. The Cooper River Bridge Corporation, chartered in June 1926 by the owners of the Isle

*2.6 The Francis Marion Hotel, circa 1930. At the time of its construction adjacent to Marion Square in 1924, the thirteen-story hotel was and still is the tallest structure in the city and the first hotel in the city to have private baths in every room. The hotel suffered significant damage from Hurricane Hugo in 1989. A renovation in the mid-1990s restored the hotel to its former glory.* Collection of Pamela Gabriel

*2.7 Abandoned gas station at the corner of Ashley Avenue and Wentworth Drive. In order to accommodate automobiles, Charleston's city streets were widened. Many old structures were destroyed to make room for service stations like the one pictured above. Often, brick from the original structures was reused in the new buildings. The stately and historic Manigault house on Meeting Street was nearly destroyed when a service station was built on the corner of its property. Most of the service stations themselves have since been destroyed.* Photograph by Jill Saxton

*2.8 King Street in the 1920s. The arrival of the automobile changed the look of King Street, the city's main business thoroughfare; two-way automobile traffic, trolley service, and parking clogged the narrow street.* Collection of Pamela Gabriel

of Palms, initiated the necessary steps to finance and build a bridge across the Cooper River. Charles R. Allen and Harry Barkerding headed the group, which had an initial bankroll of $500,000. As the promoters solicited cost estimates and government approval, the State of South Carolina and Charleston County floated the possibility of a government-sponsored bridge, encouraged by the success of their recent efforts with the new bridge over the Ashley River and the automobile bridge over Cove Inlet in Mount Pleasant. While state and local officials considered the feasibility of a public bridge, Allen and Barkerding lobbied the state to obtain a franchise for a private bridge. Such a franchise would grant the Cooper River Bridge Corporation an exclusive bridge monopoly. Without such government-sanctioned security, constructing a private bridge would be too great a financial risk. A competing bridge could sap necessary revenue away from the enterprise.

At the same time, Allen and Barkerding sought public approval for their bridge design, which was similar in plan to James Sottile's design of 1913, excepting that the new bridge would carry cars, not trains, across the harbor. Though the question of harbor accessibility was still a factor, and there were grumblings from shipping interests, Allen and Barkerding met with less hostility than did Sottile. The new bridge promoters aggressively courted city officials and business leaders and conducted public forums to discuss their plans. Furthermore, the Cooper River Bridge Corporation benefited both from a progressive atmosphere that surrounded Charleston in the 1920s and from a general change in public attitudes towards bridge construction.

Like the rest of America, Charleston was caught up in the car craze. Nationally the number of automobiles in use had nearly tripled between 1920 and 1930. The mode of travel for Americans had drastically changed, and in Charleston the need for a bridge had become acute. In 1920 there were 3,500 automobiles registered in Charleston, but by 1925 there were nearly 6,400. The last privately owned Charleston Harbor ferry service went bankrupt in 1924, causing Charleston to become even more aware of how dependent it had been on ferry transport. At about the same time tourism to the Isle of Palms reached a tremendous level. Visitors could no longer rely upon trolley service for transport to the beach, as the service was discontinued by the mid-1920s. Automobiles were now the best means of traveling between Mount Pleasant and the Isle of Palms. The general consensus in Charleston was that a bridge should be built, but debate centered on whether the new bridge would be privately or publicly owned.

*2.9 Postcard showing Charleston Harbor in the early 1900s, prior to the construction of the Cooper River Bridge. Some citizens, including Mayor Thomas Stoney, worried that the bridge would destroy the picturesque Charleston skyline.* Collection of Pamela Gabriel

*2.10 Early Charleston tour book, circa 1920. Despite the fact that many Charlestonians did not want their city turned into a "tourist town," the city found ways to make money from the visitors, including the printing of books and pamphlets for sale.* Collection of Pamela Gabriel

## The Great Debate: A Private or Public Bridge

Today most bridges in the United States are publicly maintained, and the idea of private individuals building and owning a bridge seems somewhat alien. Historically, however, the construction and control of bridges fell on private hands, a tradition dating back to the practice of levying tolls at ferry crossings and continuing when bridges replaced ferry services. In the nineteenth and early twentieth centuries, railroad companies were the principal bridge builders. The major impact of the automobile affected the way Americans built not only bridges, but also highways and streets, which necessitated redesign to accommodate motorized vehicular traffic.

Before the development of federal and state highway departments, private bridge companies were viewed as public utilities, receiving franchises that assured them a monopoly, just as utilities currently possess unique ownership of electric and telephone service. Private companies had quicker access to funds for expensive bridge projects. But as more and more automobiles crowded the streets of America, a national debate erupted over the issue of private bridges. Opponents argued that private bridge companies were unchecked and that motorists were at the mercy of the tolls levied by the bridge owners. Some feared that public road projects would be corrupted to serve the interests of private bridge companies. Proponents of private bridges noted that the bulk of toll bridges were sizable structures over major rivers and lakes and were vital conduits that promoted travel and development. They argued that the government lacked the resources and initiative for such undertakings. And finally, the definitive rejoinder was voiced—toll bridges taxed only those who used them. By the late 1920s, many bridge franchises included provisions granting government agencies the right to purchase the bridge after its construction. The government could then operate the bridge, utilizing revenue from the tolls to repay the builders. The Great Depression effectively ended the debate over bridge ownership. A dearth of private funds and the burden of ever-increasing numbers of automobiles combined to place the responsibility for highway bridges into the hands of government.

The efforts to build the Cooper River Bridge reflected the national debate regarding toll bridges. Of major concern to Charleston residents was the fact that they could be required to pay for the bridge twice—once in the form of tolls and then again in the form of increased taxes if the county later purchased the bridge. Allen and Barkerding argued strongly for a private bridge. They drew support from the Charleston *News and Courier,* which favored a privately built bridge and encouraged the group to begin construction. The effort to construct a public bridge was spearheaded by two government agencies: the state Highway Commission and the Charleston County Sanitary and Drainage Commission, which had authority over local roads. The county had expressed an interest in building a two-million-dollar, low-level bridge, but, after a series of public forums, failed to make a firm decision on the project. Frustrated with the Sanitary Commission's inability to act quickly, Charles Allen wrote a letter to the commission that was reprinted on the front page of the *News and Courier* in July 1926: "Will you definitely state that you are not interested and that you will not build? If your decision is to endorse our project we will immediately proceed."[10]

Ten days after these words appeared on the front page of the paper, Allen and Barkerding met with members of the Sanitary Commission and local representatives to the state legislature to agree upon a bridge plan. In the spirit of compromise, certain concessions were made. The public officials agreed to support the Cooper River Bridge Corporation's drive to obtain a franchise, and the county reserved the right to purchase the bridge ten years after construction. A tentative schedule of rates was proposed and agreed upon. Following the local meetings the bridge agreement was submitted to the state legislature for approval. However, during the course of these negotiations, local opposition to the low-level bridge plan had grown considerably.

Allen and Barkerding, working with engineers, estimated that the cost for a low-level bridge would be $3 million. Once it became apparent that the bridge promoters were sincere in their efforts, Charleston's powerful shipping interests began to decry the adverse impacts a low-level bridge would have on the port. As opposition to a low-level bridge grew, the promoters despaired that a high-level bridge would drive up costs, dooming their project. Then, according to John Grace, who would later become involved in the project, Harry Barkerding had a "vision" of a tunnel-bridge across the Cooper River. Whether or not the revelation was divine inspiration or an act of desperation, the bridge promoters seized upon the idea of a tunnel-bridge combination. This plan would leave shipping lanes open and still be affordable, or so the promoters thought.

The combination tube, tunnel, and bridge was slated to begin at a convenient location on the "Charleston waterfront, pass under the [shipping] channel to Castle Pinckney and thence span the [Cooper] river to the eastern end of Hog Island [present-day Patriot's Point in Mount Pleasant] about 3,000 feet below the ferry terminals and thence [by] causeway to Mount Pleasant."[11] It was this revised plan calling for a tunnel-bridge that the state legislature received and approved with certain alterations. The bill, as passed, granted the Cooper River Bridge Corporation an exclusive franchise for thirty-five years, during which time no other franchise would be granted, nor would any state agency be allowed to undertake to "build or operate a toll tube, tunnel, and bridge across the Cooper River . . . within twenty miles of Battery Point."[12] It was further agreed that the structure could not interfere with navigation in the Charleston harbor, and, finally, the project must have the approval of the United States Army Corps of Engineers, the South Carolina Highway Department, and Charleston County officials. In return, the promoters were permitted the right to assess tolls along with the right to condemn property for the necessary rights of way. The bridge company conceded to the state the option to purchase the structure any time after twenty years from the date of construction, an extension of the ten-year repurchase period originally agreed to by the bridge promoters and the county.

While the franchise bill was pending in the legislature, the promoters sought financial backing and ordered engineering feasibility studies for their project. To the surprise of the bridge promoters, the final results of the engineering study stated that a combination tunnel and bridge would be cost-prohibitive. Allen and Barkerding were now thrown into a quandary. A tunnel would be too expensive, yet they had received approval for a combination structure, consisting of tube, tunnel, *and* bridge. They quickly sought to have "or" inserted for "and" in the original franchise legislation, and counted on the legislature to act quickly on this seemingly routine matter. The South Carolina General Assembly, then at the end of its legislative session and busy with other matters, balked and insisted that the original wording remain intact. The stunned bridge promoters, vowing to "take the matter to the Supreme Court of South Carolina," solicited the help of an attorney, Charleston's former mayor John P. Grace.[13]

## CHAPTER THREE

# JOHN PATRICK GRACE

John Patrick Grace served as mayor of Charleston from 1911 to 1915 and again from 1919 to 1923. During his long political career Grace emerged as one of Charleston's leading reformers and a champion of progressive causes. He simultaneously offended many Charlestonians with his fiery temperament and inspired fierce loyalty among his supporters. Grace played a central role in developing the Cooper River Bridge, and in 1943 the bridge was renamed the John P. Grace Memorial Bridge as a tribute to his work on the project.

## Charleston Roots

Grace was a second-generation Charlestonian. His paternal grandfather, Patrick Grace, an Irish immigrant, achieved moderate success as a foreman in a rice mill and managed to send his son James to the College of Charleston. Both men fought for the Confederacy during the War Between the States. During the war, Patrick died by drowning in the Charleston harbor. James later became a bookkeeper. He married Elizabeth Daly of Charleston, and together they had ten children, six sons and four daughters. John Patrick Grace was born on December 30, 1874, the fourth child born in as many years. The family resided at 10 Society Street (since renumbered as 34 Society Street) in the Ansonborough neighborhood. The area, known as "the Borough," was populated by families of Irish and German ancestry. The Grace children attended a nearby Catholic school, Christian Brothers Academy, located on George Street. The earthquake of 1886 damaged the structure, and the school was closed by the bishop and later demolished. Young John Grace transferred to the all-male city high school located in the old Judge Mitchell King Mansion on the corner of George and Meeting Streets. The school had a reputation as being one of the best public schools in the state.[1]

The economic status of the Grace family steadily worsened following the death of James Grace in 1882. At the age of fourteen, John Patrick was forced to leave school to lend financial support to his family. For the next several years he moved from city to city, working and learning. For a time he worked as an office boy for Charleston's largest wholesale grocery firm, F. W. Wagener & Co. He then moved to Greenville, S.C., and worked with his uncle, a cotton broker. Two years later he returned to

*3.1 Photograph of John Patrick Grace taken in 1929. Grace, a colorful leader in early 1900s Charleston politics, served as mayor of the city for two non-consecutive terms. Like the bridge named in his memory, John Patrick Grace evoked strong emotions—people either loved him or hated him.*
Courtesy of Jason Annan

Charleston and held various jobs before deciding to try his luck as a steamship clerk in New York City.

During his time in New York, Grace attended meetings at the Cooper Union, where members of the Labor Party regularly debated socialism and other leftist concerns of the young labor movement. After two years, young Grace left the noise and crowds of the city and headed west. He traveled throughout the states of Indiana, Ohio, and Michigan, selling encyclopedias and absorbing the theories of social reform as proposed by the Progressive Party. The effects of these travels had a profound influence on Grace, shaping his future policies on political economy, especially government ownership of utilities. In his future political endeavors Grace would embrace the Progressive creed calling for the end of influence and privileges for special-interest groups. Armed with new insights for a better world, Grace returned to Charleston in 1896.

Despite the lack of a formal education, Grace continued his studies on his own. He eventually organized his own oil company in opposition to Standard Oil, and, while making deliveries, he would stop under the street lamps to study law. In an odd twist of fate, the ambitious young man caught the attention of a member of the Charleston elite, U.S. Representative William Elliott, a former Confederate colonel, who summoned him to Washington to serve as his personal secretary. Grace agreed to the position with the understanding that he would be permitted to continue to study law at Georgetown University.

After leaving law school in 1902, Grace returned to South Carolina to assist with Colonel Elliott's reelection.[2] He quickly became immersed in the hotbed of Charleston politics and jumped into the tumult with an unsuccessful bid for the state senate in 1902. Soon thereafter, he established a law practice with W. Turner Logan, an unlikely ally as Logan was a member of Charleston's aristocracy. Their partnership, Logan and Grace, had offices at various Broad Street addresses. Later, John Cosgrove joined the firm, and for the next thirty years, their association enjoyed a reputation as one of South Carolina's most successful law firms.

## Grace in Public Life

Traditionally, Charleston politics were the domain of the privileged, those Charlestonians known as blue bloods or Bourbons. Born into wealth, good breeding, and influence, they lived in the right neighborhoods, attended the right churches, and belonged to the most exclusive of social clubs. The "boni," as Grace termed them, controlled the banks and law firms along Broad Street and, in turn, Charleston politics. Grace's background placed him outside the sphere of this influential group, but he was determined to break its political stronghold. In his annual mayoral reports, Grace hurled accusations against those unnamed enemies who snubbed him or sought to discredit and embarrass him. Not all the slights were imagined. Following Grace's successful election to the mayor's office in 1911, outgoing mayor R. Goodwyn Rhett, a Bourbon, elected to forgo the traditional passing of keys to City Hall and chose instead to have them delivered to Grace by the building's janitor. Grace denounced this act as an insult to the citizens of Charleston.[3]

Grace was ardently anti-British; he championed a free Ireland his entire life. He abhorred the institution of slavery and, despite both his grandfather's and father's service in the Confederacy, he believed the South had received its "comeuppance" for its part in prolonging slavery. He alleged that the Civil War was fought to insure a continuance of a lifestyle for a privileged few and blamed the Bourbons for the city's postwar squalor. Grace's political modus operandi was to unite Irish and German laborers, merchants, and artisans in a block large enough to outvote the conservative political establishment. In his first annual report in the 1911 *Charleston City Yearbook*, he affirmed his beliefs: "[The] old crowd held a dying grip to the government of a city which for two centuries their policies held chained to the past. With all of Charleston's natural advantages, it should be a glorious city. Why is it not? The answer is 'Bourbonism.'"[4]

The term "fighting Irish" typified John Grace. From his first bid for political office and for the next forty years until his death in 1940, there was hardly a city, state or county election in which he was not involved. Three of the four primary elections in which Grace was a mayoral candidate were the most violent in Charleston history. During the elections of 1915, 1919, and 1923, the governor sent the state militia into

*3.2 John Grace's home at 174 Broad Street.* Photograph by Bill Morgan, Charleston, S.C.

the city to police the polls. The recount of the contested 1915 primary was marked by the accidental shooting death of reporter Sidney Cohen of the Charleston *Evening Post*. As shots rang out during the recount, ballots from a ward that strongly favored Grace were literally thrown out of a window. When the smoke cleared, Grace had lost the election to former mayor Tristam Hyde by eighteen votes. An infuriated Grace vowed to go to the courts. Succumbing to advice of friends, and in the spirit of party unity, he begrudgingly accepted the result.

The feisty Irishman was always in the thick of the turbulence that was the keystone of Charleston Democratic politics in the 1920s, even engaging in fistfights with those who disagreed with him. Grace himself described an incident that occurred during one of his campaign stumps: "Finally one man shouted, 'Grace is a crook.' I answered 'you are a liar.' The result was a riot and everybody in the hall was at once mixed up in a fight and pistols seemed to fly out of everyone's pocket. The police were called out but fortunately no one was hurt—that is, seriously hurt. I was struck in the head slightly and when I came out my coat was torn or cut, I don't know which, but in the meantime, I had done some striking myself."[5] His critics used the term "bossism" to describe Grace's political style. Bribery and fraud were commonplace. Grace himself admitted to buying votes and explained that "no man who holds high office can truthfully say he did not know that fraud was not being practiced to accomplish his election."[6]

The fact that Charleston politics was so impassioned during the early 1900s is even more surprising in light of the fact that all of the city leaders were members of the same political party—the white Democratic Party. The Republican Party, not a factor in South Carolina politics in the 1920s, was the party of Reconstruction. Dr. Henry Alexander White, a professor at Columbia Theological Seminary, tersely described South Carolina's sentiments at the time toward the Republican Party in South Carolina in his 1914 text, *The Making of South Carolina:* "Lincoln was elected by that party in the North that was most unfriendly and unfair to the South."[7] Election day in Charleston was a formality; the real outcome was determined in the Democratic primary, which excluded the city's blacks. Conservative and progressive Democrats battled each other violently before the primaries, but united on election day. In the municipal election of 1911, 1,557 voters cast their ballots. John Grace received 1,556. The one dissenter was probably not a Republican, just anti-Grace.[8]

Even Grace's detractors would have to admit to his fierce determination. He lost his first election effort for the state senate. Two years later, in 1904, he lost an election for county sheriff. In 1908 he entered the race for the U.S. Senate; again he was defeated. His first taste of victory came in 1911. Gathering strong support from the Irish, German, and labor factions of Charleston, he defeated incumbent mayor Tristam T. Hyde in the Democratic primary by a narrow majority. Four years later, following the violent 1915 election, Hyde was back in office, but Grace again challenged Hyde in 1919. The primary that year was very controversial, partly due to Grace's strong anti-British sentiments, which he passionately expounded in his newspaper, the *Charleston American.* His powerful editorials blasted both President Woodrow Wilson and Great Britain. Grace's editorial rhetoric angered many, and the U.S. Postmaster General threatened to withdraw the paper's mailing privileges unless Grace relinquished editorship of the *American.* His Anglophobic stance bordered on political suicide. The Charleston naval base employed many of those whose endorsement Grace courted, and the war preparations also gave Charleston a much-needed economic boost. When the votes were tallied, Hyde emerged the victor by one vote. Grace demanded a review of challenged votes, and the Grace-favored elections committee complied declaring Grace the winner.

Once in city hall, Grace focused his energy on upgrading the city's living conditions. During Grace's two administrations, significant efforts were made to improve the city's public health. Grace banned cows and other livestock from being kept within the city limits. During his first term, he began the initiatives that eventually led the city to develop a permanent clean water supply and citywide sewerage. Grace made Charleston City High School and the College of Charleston free for all white males. He passed modest initiatives to control Charleston's infamous bars, but his policies towards brothels were not always clear-cut. When approached by women of the

Memminger Home and School Association concerned about the "tenderloin district" that operated around Archdale, Beresford, and West streets, Grace responded that it was not his duty to suppress social evil.[9] He opposed their efforts to move the red-light district for fear that it would affect property values in other areas.[10]

Though he was against Sottile's 1913 bridge proposal, Grace was involved in improving the city's infrastructure, especially during his second mayoral administration (1919–1923). Grace was instrumental in helping to secure state and federal funding to construct the Memorial Bridge over the Ashley River (completed in 1926). During his administrations, Grace increased the number of paved city streets and by 1929, 52 miles of the city's 74 miles of streets were paved with asphalt. Grace also advanced the city tourism by improving city sanitation. His administration provided city land to construct the Fort Sumter Hotel near White Point Gardens and facilitated the construction of the Francis Marion Hotel on King Street. Grace was one of Charleston's most ambitious progressives, and he is credited for laying the foundations for modern Charleston.

## The Burden of a Lawyer

In 1923 Grace again sought re-election as mayor. This time he was pitted against a formidable newcomer to the Charleston political scene. Thomas Stoney belonged to Charleston's prestigious elite, but, like Grace, he had a flair for dramatic oratory and had learned how to court the merchants and laborers of the city. Grace was defeated in the primary, his last bid for mayor. Perhaps his tough Irish ways and notorious political style did not endear him to the newly enfranchised women voters of the city. He did not lose graciously. Historian Robert Rosen relates that a letter was delivered to Joseph P. Riley upon his election to City Hall in 1975. Addressed to "The Next Irish Mayor" and delivered by the Catholic bishop of Charleston, the half-century-old letter read, "Get the Stoneys."[11]

Following Stoney's victory, Grace settled into private law practice. In 1924, he decided to invest in the Florida land boom and, with his law partners and other prominent Charlestonians, formed the Cadillac Investment Company. Unfortunately, shortly after Grace arrived in Florida in 1925, property values declined following a series of hurricanes, and he was forced to return to Charleston late in 1926, on the verge of bankruptcy. He found the city engrossed in the bridge controversy. When Grace joined forces with those promoting the bridging of the Cooper River, he was reminded of his opposition to James Sottile's bridge proposal in 1913. He justified his shift in opinion by declaring that the new plans were visionary, while the earlier plans exploited the harbor and were proposed by "men . . . willing to furnish the money [with] nothing in view but profit."[12] He viewed the goals and ambitions of the new owners of Isle of Palms as having a higher purpose in mind, and he eagerly lent his support to their cause.

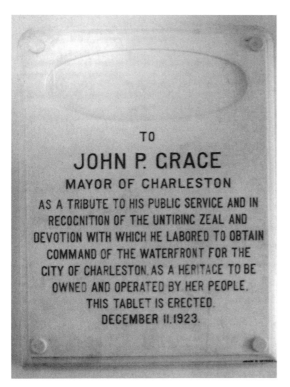

TO
JOHN P. GRACE
MAYOR OF CHARLESTON
AS A TRIBUTE TO HIS PUBLIC SERVICE AND IN
RECOGNITION OF THE UNTIRING ZEAL AND
DEVOTION WITH WHICH HE LABORED TO OBTAIN
COMMAND OF THE WATERFRONT FOR THE
CITY OF CHARLESTON, AS A HERITAGE TO BE
OWNED AND OPERATED BY HER PEOPLE.
THIS TABLET IS ERECTED.
DECEMBER 11, 1923.

*3.3 Memorial tablet to Mayor John Grace, located in Charleston City Hall. The memorial was presented to Mayor Grace on December 11, 1923, from the grateful citizens of Charleston for his "untiring zeal and devotion" in securing the waterfront property along the Cooper River from the East Shore Terminal Company. While under railroad control, the decaying docks, wharves, and warehouses in this area were an eyesore and disgrace to the city. This memorial hung in Council Chambers of Charleston City Hall until 1938, when it was removed to the Ports Utility Commission while City Hall underwent repairs. The memorial was later returned to City Hall and stored in the basement until it caught the attention of Mayor Joseph P. Riley—like Grace, a Charlestonian of Irish descent—in 1977. It was reinstalled outside of council chambers in 1980. The oval portion at the top of the tablet contains an engraving, now nearly worn away, of the city's waterfront.* Photograph by Bill Morgan, Charleston, S.C.

In John Grace's published account of the Cooper River Bridge, he credited four individuals with the success of the bridge from conception to completion: Harry Barkerding; Charles Robinson Allen; his brother-in-law, Frank Sullivan; and himself. Despite his claim Grace does not appear to have played even a minor role on the bridge project until after he was recruited to help secure the bridge franchise legislation. In reality, Grace was mired in his disastrous Florida land scheme until 1926 and did not participate in the founding of the Cooper River Bridge Company. Nor does Grace appear in any of the reports of the bridge negotiations between the company and Charleston County. Furthermore, Grace had even sought to distance himself from the impossible tube-tunnel scheme, stating that he had "nothing to do with this mistep [*sic*]."[13] Grace was certainly not among the original promoters of the bridge, but in his characteristic bravado, Grace noted that to save the project he "assumed the burden of a lawyer."[14]

John Grace was a good lawyer, and utilizing his legal talents he manipulated, influenced, and contrived with the governor, the legislature, the courts, the highway department, and financial backers to lay the groundwork for the bridge's construction. The legislature balked when asked to reword the franchise bill to eliminate the necessity of a tunnel by having "or" substituted for "and." Furthermore, the bill as passed would require construction to begin within ninety days of enactment. This was impossible for the bridge promoters as they lacked a design, a definite location, and financial backing for their project. The bill passed in the final three days of the legislative session; Grace pressured and cajoled his friend Governor John Richards to use the governor's prerogative to postpone signing the bill until after the legislature's summer recess, affording the bridge promoters additional time.

But the legal haggling was not over. The bridge project still hinged on a favorable decision by the state supreme court to ensure the validity of the franchise in light of the design changes. In December 1927, the court unanimously agreed that the franchise was legal regardless of the type of design. Grace immodestly explained his success before the judicial body: "I had not proceeded but a short way in my argument when the Court interrupted me and said that my position was so clear that no further argument was necessary."[15]

Other obstacles still had to be overcome—"a traffic survey favorable to the economic soundness of the [bridge] plan" was called for, as well as a promise that the cost of the bridge would not exceed three million dollars.[16] A traffic survey appears never to have overly concerned the bridge promoters. At the time, the harbor ferries were carrying significant numbers of passengers, and the bridge promoters most likely assumed that ferry demand would translate into a demand for a bridge crossing. They gave only passing consideration to the subject, namely by employing Grace's political muscle. Grace, in his capacity as an appointed member of the influential South Carolina Highway Commission, was charged to develop State Route 40 from Charleston to Georgetown. A bridge would facilitate this aim. The bridge financiers did secure an independent engineering estimate of the projected traffic and toll revenues—estimates made using traffic numbers from the harbor ferries. In later years, the bridge's financial troubles would prove the estimates to be far too generous.

With a bridge-tunnel ruled out by financial constraints, the only option left to the bridge promoters was a high-level span. Allen and Barkerding had initially, and wrongly, estimated that the cost of a high-level bridge would be the same as a low-level structure, $3 million. Their original financial connections in New York and in Milwaukee had been severed and new backers had to be found. Utilizing the influence of his brother-in-law, Frank Sullivan, Grace turned to Chicago and made contacts with

*3.4 John Grace's gravesite in St. Lawrence Cemetery, Charleston. Grace died in 1940. The inscription on the tombstone, a quote from Judge William Brawley, refers to Grace's successful lawsuit to overturn South Carolina's system of peonage.* Photograph by Bill Morgan, Charleston, S.C.

both Federal Securities and the H. M. Byllesby Company. In a July 1927 agreement, Federal Securities agreed to finance up to $3 million of the bridge costs in return for 77.5 percent of the stock in the bridge corporation.[17] When estimated costs for the high-level bridge jumped to $6 million, negotiators for the bridge prevailed upon both Federal Securities and the H. M. Byllesby Company to guarantee completion of the bridge across the Cooper River despite the increased costs.[18] With the franchise bill approved and the funds secured, the grateful bridge promoters named John Grace president of Cooper River Bridge, Inc. Other officers were vice presidents J. J. Shinners and Harry Barkerding, secretary W. G. Pohl, assistant secretary and treasurer Charles R. Allen, and general manager C. A. Miller. What followed were a series of arm-twisting negotiations that would determine the location of the bridge, and ultimately, transform the entire appearance of the Charleston Harbor.

## CHAPTER FOUR

# BUILDING THE BRIDGE

## Selecting a Location

The ink was barely dry on the financial agreements when negotiations to select the location of the bridge nearly doomed the entire Cooper River Bridge project. All planning had assumed that the bridge would begin at "some convenient point" on the Charleston peninsula, presumably near the center of town.[1] In 1913 James Sottile had proposed connecting his bridge at the foot of Calhoun Street, near busy King and Meeting Streets. Like Sottile's plan, the Cooper River Bridge Corporation's original low-level bridge proposal was devised so that it would cross the harbor at its narrowest point. After securing a bridge franchise from the state, and after agreeing to a high-level bridge, the promoters targeted Market Street as the launching point for the Cooper River Bridge. Market Street's proximity to Charleston's main streets (Broad, King, and Meeting) made it part of what the promoters called "the center of the locally owned automobile territory."[2] Market Street's location also accommodated the movement of construction materials and gave vehicles easy access to the construction site. Significantly, no buildings would need to be destroyed or removed along this route.

The Market Street approach initially had strong public support as well as backing from Charleston's city council, yet there remained one influential opponent: the city's Port Utilities Commission. The commission was adamantly opposed to bridge support structures on its property at Union Pier. In a letter to the War Department (which had jurisdiction over the matter by virtue of the naval shipyard in Charleston), Port Utilities commissioner Frederick Davies stated his opposition to a bridge at the mouth of the harbor. He wrote, "If a bridge is constructed at the very gateway of our harbor, it will immediately place the port at a disadvantage."[3] Davies proposed a new location, one mile further up the Charleston peninsula. In a rebuttal letter sent to Davies, the Cooper River Bridge Corporation described the new location as "far too remote from the city. It would encourage tourists to by-pass the business district . . . and [the bridge] would necessarily be much longer and far more costly than the one at Market Street."[4] The Ports Utilities Commission remained immovable.

4.1 *Illustration from 1926 of the proposed Market Street approach to the Cooper River Bridge. Early bridge plans called for the span to cross the harbor near the U.S. Customs House on East Bay Street. Two approach ramps were designed to start at Meeting Street and would have run along either side of Market Street. At Church Street the two ramps would join into a single raised roadway, passing beside the U.S. Customs House. The causeway would then turn and cross Concord Street, cutting through Union Pier, the city-owned port. This plan would have created an elevated roadway over much of the market, an area that became the heart of the city's tourist industry; it was abandoned after fervent protests from Charleston's shipping interests.* Courtesy of the U.S. Customs Office, Charleston, S.C.

In October 1927 the War Department held a public hearing to determine the merit of the Market Street approach. During the meeting both businessmen and representatives of Charleston's civic organizations expounded the virtues of the proposed approach site. The *News and Courier*, a staunch supporter of a Cooper River bridge and its promoters, featured an editorial praising the plan, complete with an artist's rendering depicting the favorable effect of the bridge alongside the U.S. Customs House.

For two months, as the War Department considered the case, Grace and his associates fretted. Fearing the worst, Grace attempted to protect his reputation. He wrote to bridge financier J. J. Shiners of the H. M. Byllesby Company, "I feel that all my promises in the matter have been performed 100% . . . I never promised that I would be able to steer the matter through the War Department."[5]

The War Department's final decision was bad news for the bridge promoters. The government opposed the Market Street approach and demanded that any new bridge have a horizontal clearance of 1,800 feet to allow passage of naval vessels. The decision created an emergency for both the Cooper River Bridge Corporation and its design engineer, Shortridge Hardesty. The company could not afford a bridge of this size. Upon hearing of the War Department's requirement, Hardesty remarked to Grace, "That means no bridge. Unless the recommendation can be modified very substantially, you will have no bridge across the Cooper River."[6] Surprisingly, the War Department's verdict was not the fatal blow the bridge promoters expected. Hardesty and Grace were able to prevail upon the officials of the War Department for a compromise. The promoters agreed to move the location of the bridge to a site desired by the government. The new bridge would cross Drum Island, one mile distant from the Market Street location and away from the city's piers. Various streets were presented as takeoff points: Sumter, Lee, and an unnamed street just north of the Columbus Street shipping terminals. In turn, the War Department agreed to reduce the required horizontal clearances to 1,000 feet, though this still represented an increase from the 850-foot span originally proposed by the Cooper River Bridge Corporation. Vertical clearance was set at 150 feet above the high-tide water level. The political battle was won, the financial backing secure, and the final hurdle—the bridge design—began in earnest.

## Waddell and Hardesty, Bridge Engineers

While securing a franchise for their bridge, the Cooper River Bridge Corporation contracted the engineering firm of Waddell and Hardesty, of New York, to begin preliminary engineering work. At the time the firm was enjoying national recognition for its bridge designs, especially those in New York City. Frank Sullivan, Grace's brother-in-law, was an engineer living in New York, and it is likely that he introduced the bridge promoters to Waddell and Hardesty. The firm was internationally known for its designs of movable bridge spans, such as drawbridges, and, in fact, had engineered the nation's first working lift span. Because the Cooper River Bridge Corporation first considered a low-level bridge with movable spans, Waddell and Hardesty was an excellent choice. After the low-level plan was eventually ruled out in favor of the high-level design, the firm was nonetheless retained to design it.

The firm's founder was Dr. John Alexander Low Waddell, one of the nation's most widely respected, though controversial, bridge designers. Born in Port Hope, Ontario,

*4.2 Dr. John Alexander Low Waddell as he appeared at the height of his career. The large medallion on Waddell's coat pocket was awarded to him by the Emperor of Japan in recognition of his engineering accomplishments.* Courtesy of Dr. Henry Petroski, Duke University

Canada, in 1854, Waddell was a graduate of Rensselaer Polytechnic Institute in New York and McGill University in Canada. He established an engineering practice in Kansas City, Missouri, eventually moving his business to New York City in 1920. As was the case with many engineers of his time, Waddell's successes covered a spectrum of projects, ranging from mining to marine engineering, and he was involved in variety of significant projects around the world.[7] Waddell's engineering acumen earned him international acclaim and even a professorship at the Imperial University in Tokyo, Japan. One of Waddell's lasting, and controversial, legacies was his 1926 textbook, *Bridge Engineering*, which remains a useful reference for engineers. In writing the book, Waddell utilized existing bridge structures to illustrate both good and bad design techniques. Waddell harshly critiqued the designs of his colleagues and often presented his own projects as illustrations of exemplary engineering practices. Waddell's egocentric and self-congratulatory attitude did not endear him to his fellow engineers.

In sharp contrast to Waddell, Shortridge Hardesty is remembered by his colleagues and his family as an unassuming, religious man of even temperament. Hardesty's son, Egbert, remembered that even during periods of great stress, the elder Hardesty never uttered an oath stronger than "rats."[8] He was a Midwesterner, born in Weston, Missouri, in 1884. Like Waddell, Hardesty attended Rensselaer Polytechnic Institute,

*4.3 Shortridge Hardesty, designer of the Cooper River Bridge. On business trips, Hardesty carried several suitcases of engineering data. Using this information he was able to formulate a preliminary design of a bridge while on site, in a matter of hours.* Courtesy of Hardesty and Hannover

where he was an exemplary student. Shortly after graduation, Hardesty began his career in bridge engineering, accepting an entry-level draftsman position at Waddell's firm in Kansas City. Hardesty's engineering talent soon became apparent to Waddell. Following the death of his son and successor during the influenza epidemic of World War I, Waddell relinquished more and more responsibility to the younger Hardesty. Eventually Hardesty became chief of operations, assuming control of the firm's projects. In 1927, Hardesty was made a full partner, and he was chiefly responsible for overseeing the design of the Cooper River Bridge. Hardesty's reputation as a bridge engineer grew, and he practiced until his death in 1954. Today the American Society of Civil Engineers presents the Shortridge Hardesty Award to honor structural engineers who have made exceptional contributions to their field.

## Selecting a Bridge Design

The earliest designs for a Cooper River bridge had called for a low-level structure with mechanical spans that were raised and lowered to accommodate shipping traffic. James Sottile's unsuccessful bridge proposal of 1913 provided for several draw spans that would open to a vertical position. The publicly financed bridge briefly considered by Charleston County in 1927 would have employed the use of lift spans—spans that are raised horizontally. The original plans of the Cooper River Bridge Corporation

## 4.4 *Significant Cantilevered Truss Bridges*

| Year Completed | Name | Location | Span Length (feet) |
|---|---|---|---|
| 1917 | Quebec Bridge | Quebec, Canada | 1,800 |
| 1890 | Firth of Forth Bridge | Scotland | 1,710 |
| 1974 | Nanko Bridge | Japan | 1,673 |
| 1974 | Commodore John Barry Bridge | Chester, Pa. | 1,644 |
| 1958 | Greater New Orleans Bridge | New Orleans, La. | 1,575 |
| 1985 | Mississippi River Bridge | New Orleans, La. | 1,575 |
| 1995 | Gramercy Bridge | Gramercy, La. | 1,460 |
| 1936 | Transbay Bridge | San Francisco, Calif. | 1,400 |
| 1968 | Baton Rouge Bridge | Baton Rouge, La. | 1,235 |
| 1955 | Tappan Zee Bridge | Tarrytown, N.Y. | 1,212 |
| 1930 | Lewis and Clark Bridge | Longview, Wash. | 1,200 |
| 1976 | Patapsco River Bridge | Baltimore, Md. | 1,200 |
| 1909 | Queensboro Bridge | New York, N.Y. | 1,182 |
| 1962 | Las Americas Bridge | Panama City, Panama | 1,128 |
| 1927 | (First) Carquinez Strait Bridge | Vallejo, Calif. | 1,100 |
| 1958 | (Second) Carquinez Strait Bridge | Vallejo, Calif. | 1,100 |
| 1961 | Second Narrows Bridge | Vancouver, B.C., Canada | 1,100 |
| 1930 | Jacques Cartier Bridge | Montreal, Quebec, Canada | 1,097 |
| 1968 | Isaiah D. Hart Bridge | Jacksonville, Fla. | 1,088 |
| 1956 | Richmond– San Rafael Bridge | Richmond, Calif. | 1,070 |
| 1929 | Grace Memorial Bridge | Charleston, S.C. | 1,000 |
| 1980 | Newburgh-Beacon Bridge | Newburgh, N.Y. | 1,000 |

also favored the use of mechanical spans. Critics of this design rightly argued that the massive motors required for raising and lowering mechanical spans were prone to failure. Furthermore, no draw or lift span could accommodate the 1,000-foot horizontal clearances required by the War Department in 1927.

When the bridge location and clearance requirements were made firm, the design process began in earnest. The challenge for Hardesty was to design a structure that incorporated two spans, a smaller one over Town Creek (on the Charleston side of Drum Island) and a larger, 1,000-foot span over the Cooper River that would provide a 150-foot vertical clearance, for a price not to exceed $6 million (the amount Grace had secured from his Chicago investors). Perhaps the most significant design factor limiting Hardesty was cost. Grace noted that Hardesty "had been given a dead line [*sic*] on cost," and cost dictated that Hardesty opt for a functional, utilitarian bridge design.[9]

The decades between 1910 and 1930 were a period of tremendous road improvement and bridge construction nationwide. Across the country, bridge engineers were designing structures that were architectural masterpieces. Novel designs were engineered and perfected in this era, including the suspension bridge and the steel arch. Financed by the ample coffers of large industry, Othmar Ammann and Gustav Lindenthal completed the massive, art-deco style Hell Gate Bridge, a steel arch spanning New York's East River, in 1916, at a cost of $20 million. During the late 1920s Ammann began an even more magnificent arch, the Bayonne Bridge in New York City, a $13 million, publicly financed project. During the same era the monumental Golden Gate Bridge in San Francisco was being designed. After four years of construction this publicly financed suspension bridge was completed in 1937 at a cost of $45 million. By comparison, the Cooper River Bridge Company, without support from industry or government, struggled to finance its proposed $6 million bridge. In short, the Cooper River Bridge Company was hindered by financial constraints, and Hardesty could not afford to experiment with novel designs or excessive architectural features. In addition to being prohibitively expensive, other bridge designs like a steel arch or suspension bridge were ill-suited for Charleston's soft, sandy soil. Although there is no record of any alternative bridge designs Hardesty may have considered for the Cooper River Bridge, he would have recognized that under the clearance requirements, location, and financial constraints only one design was suitable: a cantilevered truss.

By the late 1920s, the cantilevered truss had been employed in several high-profile bridge designs and had a proven utility. For example, the Quebec Bridge over the St. Lawrence River in Canada, a railroad bridge completed in 1917, had a clear span of 1,800 feet (as of this writing it is still the world's longest cantilevered truss bridge). The elegant Queensboro Bridge over the East River in New York City, mentioned in

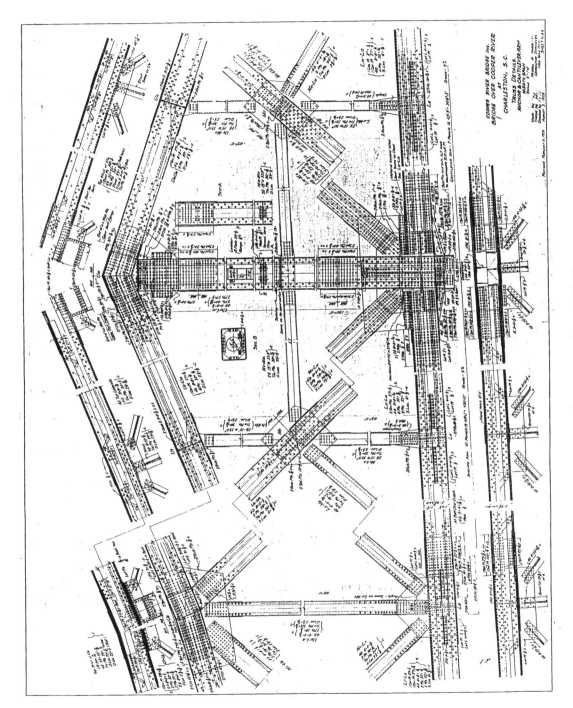

4.5 *Page from the original blueprints for the Cooper River Bridge.* Collection of Pamela Gabriel

F. Scott Fitzgerald's *The Great Gatsby*, was completed in 1909 and spanned 1,182 feet. At the time of its construction the Cooper River Bridge (specifically the Cooper River span) was the world's fifth longest cantilevered truss, with a horizontal clearance of 1,000 feet. Compared to other bridges built during the same era, the Cooper River Bridge was not a staggering feat of engineering or an architectural masterpiece. However, it was an unusual, if not exceptional, bridge design and attracted international attention.

The final design of the Cooper River Bridge was begun in the fall of 1927 and was completed by the spring of 1928. The design process started with Shortridge Hardesty. Hardesty, perhaps assisted by a senior engineer, developed the conceptual bridge design by selecting the cantilevered construction, finalizing the bridge route, and formulating a rough estimate of the design of the bridge structure. Hardesty was responsible for the shape and form of the bridge in much the same way that a senior architect is responsible for the outward appearance of a building. Despite restrictions on costs, Hardesty was charged not only to design a high-level bridge to cross the Cooper River, but also to engineer a structure that would preserve the natural beauty of Charleston Harbor, a condition that Mayor Thomas Stoney made very clear in 1926.[10]

Indeed Hardesty wore two hats: engineer and artist. The engineer must possess the technical knowledge to calculate the strength of steel and concrete components that support the structure. Hardesty's team of assistants calculated a myriad of stresses and strains, all by hand. Equally important, the engineer as designer must take the steel and concrete and generate a structure that is both safe and visually appealing. The Cooper River Bridge would be the tallest structure in Charleston; it would be a steel rainbow over the harbor and a backdrop for Charleston's spires and rooftops. Charleston's leaders looked to Hardesty to create a structure that would function as a gateway befitting their revitalized and beautiful city.

Junior engineers used Hardesty's conceptual design and developed a detailed bridge design. The detailed design process was time- and labor-intensive. Engineers specified every component of the bridge, down to the number and size of the steel rivets. After extensive reviewing and refining, junior engineers drew their final designs on paper. Draftsmen, many of whom were actually engineers fresh out of school, then traced the final drawings onto linen sheets. Blueprints could be reproduced from the linen drawings. The drawings of the Cooper River Bridge are masterful art: finely drawn trusses, cross-sections, and concrete piers, completely detailed and shown in various perspectives.

After Hardesty and his engineers completed the design, the Cooper River Bridge Corporation solicited bids from contractors. Grace explained, "Mr. Hardesty was at the time extremely pessimistic."[11] So tight was the project's budget that both men secretly believed that no construction company could build the bridge within the limits of the project's resources. But when contractors submitted their final bids, Hardesty was pleasantly surprised. Grace later dramatically described the anxiety of the moment: "We sat around the table in fear and trembling. We knew that all our efforts, all expenses that . . . were already out of pocket . . . would be kissed goodbye and the Cooper Bridge would be forgotten unless for the last time luck was with us. When the bids were opened we were genuinely . . . surprised."[12] The bids came in under the cost limit and the project was on.

## A Cantilevered Truss

The zigzag configuration of steel and concrete that constitutes the form of the Cooper River Bridge may seem extraordinarily complex, but the design of the Cooper River Bridge is really quite simple. The first step to understanding the engineering of the bridge is to dissect the basic components of the bridge structure. The Cooper River Bridge's spans are actually only the portions of the bridge that cross the main shipping channels. There are two primary spans in the Cooper River Bridge: the 1,000-foot span over the Cooper River, and the 640-foot span over Town Creek. The portions of the bridge that access the spans are considered causeways, or approaches. Skeletal-looking steel trusses, frameworks of steel beams riveted together in triangular formations, cap the spans of the Cooper River Bridge and form the bridge's superstructure. The truss is a vital part of the bridge, designed to carry the weight, or loads, placed on the bridge and to support the bridge's massive size. In fact, the largest load carried by any bridge is usually its own weight, which normally exceeds that of any traffic the bridge may be required to hold. Finally, the bridge is held aloft by huge concrete piers, collectively known as the substructure, which extend deep into the harbor sediment.

The most visible parts of the bridge are its strong trusses. If the Cooper River Bridge had been constructed only of several long, steel beams spanning the river, the beams would have bent from their own weight, and snapped when a single car passed over them. The Cooper River Bridge trusses are used to transfer loads from any one point of the bridge to other parts of the bridge and eventually to the foundation via the concrete piers. Because every steel beam in a truss carries only a small part of the total load, engineers use trusses to span distances that solitary beams could not cross.

However, even trusses do not have the strength to span the longest distances. At the time the Cooper River Bridge was designed, the longest continuous truss bridge in the world was the Sciotoville Bridge over the Ohio River in Sciotoville, Ohio, built in 1917, which spanned a horizontal clearance of 775 feet, 225 feet short of the clearance required for the Cooper River Bridge. Furthermore, trusses have failed during high winds, a significant consideration in hurricane-prone Charleston. Because of the shortcomings of the truss, Hardesty had to employ a sophisticated variation known as the cantilevered truss. The design of the cantilevered truss allows engineers to extend the reach of a truss while still retaining the benefits of a truss design, namely economy. The first steel bridge to employ a cantilevered truss in its design was Scotland's famous Firth of Forth Bridge, a railroad bridge designed by British engineers Benjamin Baker and John Fowler and completed in 1890. The Firth of Forth Bridge is a monster with horizontal clearances of 1,700 feet. The first of its kind, it remains the world's second longest cantilevered truss (though cantilevered trusses are not, by far, the longest spans achievable: the Golden Gate Bridge, a suspension bridge, has a clear span of 4,200 feet).

A simple cantilever is a structure that is supported at only one end, like a diving board. Cantilevering is common in engineering, and the cantilevered truss is an application of this principle applied to bridges. The concept can be illustrated with a playground seesaw constructed of a long wooden plank and supported in the center by a fulcrum. A child sitting on one end of the seesaw can balance a child sitting on the opposite end. In this illustration, the portion of the wooden plank extending from the fulcrum to one child is called the cantilever arm. The other end of the wooden plank is known as the anchor arm. Without the weight of the child on the anchor arm, the child on the cantilever arm would fall to the ground. In fact, if the anchor arm of the

*4.6 Firth of Forth Bridge in Scotland, the first steel bridge to use a cantilevered truss. The massive bridge remains the second largest of its type in the world. Note the suspended span, seen near the center of the photograph: it appears tiny compared to the massive cantilever arms that support it.* Photograph by Martin Junius, copyright 1994–2001, http://www.m-j-s.net/photo/; used with permission

seesaw were permanently secured (by cementing it into the ground or fastening a large weight to its end, for example), a child on the end of the cantilever arm would be held aloft in a stable position. In the most basic terms, the cantilevered truss is a giant seesaw with a secured anchor arm. The Cooper River Bridge has four cantilevered trusses, a pair for each span.

The profile of the main trusses of the Cooper River Bridge features a peculiar wave shape. Each span's truss has two peaks. Between the two peaks is a part of the truss that "humps" slightly. The physical features of the trusses can be used to identify the various parts of the bridge. Each span of the Cooper River Bridge—the Cooper River span and the Town Creek span—consists of two anchor arms and two cantilever arms. The beginning of each anchor arm is the point where the overhead truss starts and stops. Here the truss ties to a special pier called the anchor pier. The highest points on the trusses are aligned with the large piers called cantilever piers. This is the point at which the anchor arm ends and the cantilever arm begins (the fulcrum in the seesaw analogy). The cantilever arms extend from the cantilever piers into the main shipping channel and form part of the bridge's main spans; however, the anchor and cantilever arms cannot freely rotate about the cantilever pier. By securing the ends of the anchor arms to piers (the anchor piers) the bridge is anchored to the sediment below harbor bottom. The massive weight of each anchor pier holds its cantilever arm upright.

Each span of the bridge has two cantilever arms that extend from opposite sides into the center of the channel. The cantilever arms do not connect. Rather, the cantilever arms are separated by the suspended span, or the portion of the bridge directly in the center of the shipping channel. The suspended span can be easily recognized as the slightly "humped" portion of the bridge truss. In reality, the suspended span is a simple truss that rests on the ends of the cantilever arms. The suspended span is connected to the ends of the cantilever arms by large hinges located at the bottom of the truss. The hinges allow the ends of the suspended span to rotate about its connection points. If the suspended span were to be removed the cantilever arms would remain in place. (When dismantling the bridge, contractors can simply drop the suspended span into the water by removing the hinges located at the base of the truss.)

The use of a suspended span may appear to defy logic, but it is a clever way of facilitating the design of the bridge. The suspended span operates much like a gangplank hooked between two ships. As the ships undulate the gangplank will pivot about its connecting hooks, yet despite the movement of the ships the gangplank stays in place. In fact, the trusses on the Cooper River Bridge do move. The weight of automobiles passing over the bridge causes the cantilever arms to deflect downward. Because the ends of the cantilever arms are hinged only to the suspended span, which rocks somewhat about its hinges, the minute deflections to one cantilever arm do not affect the

4.7 *Photograph taken shortly before the completion of the Cooper River Bridge and showing the basic components of a cantilevered-truss design. The superstructure of the Cooper River span is composed of a suspended span, two cantilever arms, and two anchor arms. The suspended span can be recognized as the slightly "humped" portion of the truss, right of center and still under construction in this photograph. Cranes stand at either end of the suspended span. The suspended span is hinged to the ends of the cantilever arms. The pointed "peaks" of the bridge's truss mark the interfaces between the cantilever arms and the anchor arms. The end of each anchor arm is the point where the overhead truss connects to an anchor pier. Courtesy of the Charleston County Library*

opposing cantilever arm. The rotational force of the suspended span about the ends of the cantilever arms is known as a *moment*. By allowing free rotation about its axis, the hinge pins prevent the transfer of moments from one span to the next. If the ends of the cantilever arms were tied directly to each other, deflections would impact the entire span, causing considerable stresses in the steel and greatly complicating the bridge's structural design.

## The Bridge Is Built

Contracts for the work on the Cooper River Bridge were divided among four major companies. The Charles E. Hillyer Company of Jacksonville, Florida, was responsible for the approaches and viaduct (an elevated roadway) that traverses Drum Island. The Foundation Company of New York was awarded the contract for the thirteen piers that support the structure. The McClintic-Marshall Company of Pennsylvania was selected to build the main spans with steel supplied by Virginia Bridge and Iron Company of Roanoke, Virginia.

A vital first step in the construction of the Cooper River Bridge was pinpointing the exact location of the piers. Even the slightest error could result in a misalignment

*4.8 Photograph taken on June 18, 1929, showing progress on the Cooper River span of the Cooper River Bridge. Two large cranes sit on the ends of the span's two cantilever arms. One of the span's anchor arms, shown in the left of the photograph, is supported by temporary scaffolding. The weight of the anchor pier, acting through the anchor arm, keeps the cantilever arm aloft.* Courtesy of the Charleston County Library

*4.9 Photograph taken on May 23, 1929, showing one of the Cooper River Bridge's many bearing shoes. The bridge's truss and road deck connect to the concrete piers via massive hinges known as bearing shoes. As traffic crosses the bridge, these hinges allow the superstructure to rotate slightly about the pier, thus minimizing excessive stress to the steel truss.* Courtesy of the Charleston County Library

of the entire structure. Surveys commenced on February 7, 1928. The process was so demanding that even the tape measure used had to be analyzed to account for its thermal expansion and contraction. Surveying the slime and mud-filled tidal marshlands along the Cooper River proved to be an ordeal. Surveyors sank knee-deep in the mud and had to be extracted by their coworkers. The resident engineer, Charles K. Allen (not to be confused with Charles R. Allen, one of the original bridge promoters), reported that he knew "of no bridge where this work was harder to accomplish."[13] To complete the surveys Allen reported that "special tripods and platforms had to be built for observation points."[14] Tripods were driven into the marsh twelve to sixteen feet before solid footing was located. Despite these handicaps the surveys for the bridge were completed with less than a two-second (1/1800 degree) error.

*4.10 Photograph taken on July 2, 1929, showing construction of the suspended span over the Cooper River. At the top of the truss, adjacent to each crane, note the small gaps in the steel. Beams have yet to link the top of the suspended span to each cantilever arm. Below these gaps, at the bottom of the truss, the suspended span and cantilever arms are joined with massive hinges. The hinges allow rotational movement, permitting the suspended span to "rock" about the ends of the cantilever arms thus minimizing stress to the steel truss.* Courtesy of the Charleston County Library

Hundreds of Charlestonians attended the official groundbreaking ceremony on May 19, 1928, to mark the beginning of construction on the Town Creek span. The first stage of construction was placing the support piers. It was necessary for the piers to rest on solid ground, as a weak footing could cause them to shift or settle. The marshlands along the Cooper River presented a problem for the foundation contractor. At certain points in the riverbed the firm clay, lime, and sand deposit, known as marl, that was to provide support for the piers lay below forty feet of water and mud. Adequate foundations for the largest piers, the anchor and cantilever piers, could only be found deeper, nearly a hundred feet below the river surface.

When constructing the shallow piers—those extending around thirty to forty feet below the water surface—contractors employed the use of a cofferdam, a temporary enclosure of wood or steel plates arranged in a circle. The cofferdam was first erected around the pier location. Once it was in place, water was pumped out of the enclosed area. The plates of the cofferdam held back the water, allowing cranes to excavate below the river bottom. After the excavation was completed, bearing piles were driven into the exposed river bottom. The long concrete or wooden piles carried the weight of the pier deep into the sturdy marl. Next, a concrete cap was poured over the piles to provide a base for the pier body.

Some of the bearing piles were wooden, and, once placed, were surrounded by mud and out of contact with oxygen. There they were expected to remain in a good state of preservation. However, during the construction of the of the second bridge over the Cooper River, the Silas N. Pearman Bridge, in 1964, engineers discovered that the older bridge had shifted nineteen inches out of alignment. Divers determined that the river current had worn away the base of one pier, exposing some of the wooden bearing piles. Woodworms had eaten away at the exposed pilings, causing the pier to shift. The damaged pier was tied to the Pearman Bridge while the support was repaired.

Though sufficient to construct most of the foundations for the Cooper River Bridge, the cofferdam method could not be used to excavate the deepest piers—the massive anchor and cantilever piers. At significant depths (over fifty feet below water) the weight of water and mud would have collapsed a cofferdam. To reach the depths required for the deepest piers the Foundation Company utilized a special digging method known as the pneumatic caisson process. James Eads first used the pneumatic caisson in the United States in 1868 during construction of his bridge across the Mississippi River at St. Louis (now known as the Eads Bridge). With the pneumatic caisson process, Eads was the first to bridge the strong, swirling currents and changing riverbed of the Mississippi, despite the reservations of his contemporaries.

The pneumatic caissons used during construction of the Cooper River Bridge were actually giant wooden frames that served as forms for the concrete support piers. Each caisson was constructed on land and then floated into the river for placement. Once in the proper spot, the form, which was equipped with steel "cutting edges" on its bottom, was hammered into the bottom of the river. Like the cofferdam, the caisson acted as a dam, keeping out mud and water. Each caisson was built with a watertight working chamber at its bottom. The harbor bottom formed the floor of the chamber, and the roof was a strong wooden ceiling that spanned the length and width of the chamber—55 feet long, 18 feet wide, and 8 feet high. When the caisson was in place, water was pumped out of the form and from the working chamber. In shifts the laborers, known as sandhogs, descended via a narrow shaft into the working chamber to dig out the mud and sand of the river bottom.[15] Excavated mud was removed to the surface via special lifts constructed inside the caisson. The excavated material was used as fill for the bridge approaches.

As the sandhogs dug, tons of concrete were poured onto the roof of the working chamber. The sturdy roof of the working chamber was designed to hold the increasing weight of the concrete, which pushed the caisson deeper into the harbor bottom, matching the pace of the excavation performed by the sandhogs. The height of the working chamber was kept constant by skillfully balancing the amount of concrete poured with the progress of the sandhogs' excavation. As the caisson sank deeper into

*4.11 Traditionally, the chief resident engineer would receive the plaudits for a successful engineering project. But the strong personality of John Grace overshadowed the role of resident engineer Charles Keyes Allen, a situation Allen later protested. Allen was born in Ohio but raised in Missouri, where he attended the Missouri State School of Mines and Missouri State University. A tribute to Allen was recorded in "The Story of the Bridge," a booklet printed by the Francis Marion Hotel for the opening of the Cooper River Bridge in 1929.*

*In 1900 he moved to Kansas City and was engaged in the drafting of designs of bridges for Dr. J. A. L. Waddell, working in that capacity for about nine months, and resigning to become Chief Inspector in the Water Department of Kansas City, Missouri. He returned to Dr. Waddell in the latter part of 1908 and has been with him most of the time since, during which period he has had the charge of the construction of several important large bridges, as well as making the surveys, borings and other investigations for obtaining the necessary information on which to base the designs of many large bridges.*

*In his work as Resident Engineer of the Cooper River Bridge, Mr. Allen has demonstrated his splendid ability as an engineer and executive. Many grave, delicate and difficult problems have arisen during the building of this large structure which have called for immediate mature decisions. These emergencies Mr. Allen has met in a masterful manner, and the fact that this gigantic structure has been completed many months before the contract penalty date, is a standing testimonial to Mr. Allen's engineering skill.*

Photograph from "The Story of the Bridge," courtesy of College of Charleston Special Collections Library

the river bottom, the concrete resisted the pressure of water and mud on the caisson walls and prevented the form from collapsing. Once the piers were dug to the proper depth the working chamber was abandoned and filled with concrete. A floating concrete-mixing plant prepared the enormous amount of concrete required for the piers. Over 80,000 tons—more than 38,000 cubic yards—of concrete were poured into the thirteen piers supporting the Cooper River Bridge.

It is an understatement to say that the sandhogs' work was harsh. The job required experienced workers, and most of the laborers employed as sandhogs came from other parts of the country where they had worked on similar projects. As the sandhogs dug into the river bottom, the excavated soil became dry; the working chamber was surprisingly water-free. However, the interior of the working chamber was dimly lit, the air pumped from the surface was dank and humid, and the work dirty and backbreaking. Pressure built up within the working chamber and increased with depth. As sandhogs returned to the surface, they risked developing caisson disease, also known as the bends, an ailment often associated with deep-sea divers. This condition occurs when the body experiences a sudden drop in pressure, causing air bubbles to accumulate in the blood. Mild cases of this condition result in cramps and pain, but severe cases can end in death. James Eads lost several sandhogs to the bends when building his bridge over the Mississippi River. To treat the effects of this condition Eads devised the first depressurization chamber, a predecessor to the modern hyperbaric chamber, which slowly acclimated workers to atmospheric pressure. A depressurization chamber was used during the construction of the Cooper River Bridge. Despite such safety precautions, one worker died from the bends during construction.

At one point during construction of the Cooper River Bridge sandhogs developed unusual symptoms, their eyes and lips swelled, causing extreme discomfort. Work was halted temporarily as engineers were stymied by the problem. The cause of the malady was found to be hydrogen sulfide gas released from the muck of the riverbed. To remedy the situation additional air was pumped into the working chamber, forcing the gases out under the edges of the caisson. Compared to other workers on the Cooper River Bridge project, the sandhogs labored in the harshest conditions. However, their working day consisted of two three-hour shifts, as opposed to the ten-hour days worked by others. During the initial months of construction, novice sandhogs were paid $6.50 per day (about $60 in year-2000 dollars). After a serious construction accident, the Foundation Company had trouble recruiting men for the harsh task of digging, and the pay for experienced sandhogs eventually soared to $14 per day (roughly $139 in year-2000 dollars).

The pneumatic caisson process was used to construct the six largest piers on the Cooper River Bridge (all four cantilever piers and the two anchor piers of the Cooper

River span). Each of these piers has a "belled" foundation—the base of the pier is wider than the body. Belling the foundation distributes the weight of the pier over a wider area, thereby minimizing compaction of the marl and making the bridge foundation more stabile. Constructiong the foundations in this manner increased the hazards of collapse, as workers were essentially digging under the caisson edges. In the risky pneumatic caisson process, danger was ever-present.

## Tragedy Strikes

December 1, 1928, was the most tragic day of construction on the Cooper River Bridge. That day, seven men died while completing one of the anchor piers (pier number ten) of the Cooper River span. As workers dug the pier, one end of the caisson hit a pocket of firm material, either oyster shells or a patch of dense sand, causing the form to sink unevenly. The tons of concrete in the form made the caisson top-heavy and it started to tilt dangerously. Initially the fifteen sandhogs working on the pier escaped safely. When the night shift of ten men and a foreman returned to right the caisson, it twisted, and tons of mud rushed into the working chamber, burying seven of the men alive. The *News and Courier* offered the following analogy to explain the accident: "It was as though a glass, inverted in a tub of water, had suddenly been tipped to allow a cushion of air, keeping out the water, to escape, and the water to take its place, filling the glass."[16]

The seven who died in the accident were James Brown, Coley Gray, Bailey Hightower, Theodore Hill, Arthur Johnson, Clifton Moore, and Albert Ross. The local newspaper noted ominously that the accident occurred during construction of pier number ten, the thirteenth pier placed, and occurred on a Friday. "Through the ages, the number and the day have been considered ill-starred, unlucky and portending evil. If such be true, the combination worked with a vengeance."[17] Despite the caisson tragedy, construction resumed. Resident engineer Charles K. Allen stoically remarked, "These catastrophes occur once in a while. They are unfortunate . . . There have been others in the past, which were even more dreadful than the one which occurred [in Charleston]."[18]

After the accident, the Foundation Company offered compensation packages to the victims' families. Two families refused compensation and opted to sue. Ironically, John Cosgrove, law partner of Cooper River Bridge Corporation president John Grace, represented the estate of both victims. One of the victims, Arthur Johnson, also known as Arthur King, not only had two names but also had two wives, as was revealed after accident, and both wanted compensation. When told of the accident, Johnson's estranged wife, Francis King (Johnson's legal wife), requested $100 for both train fare to Charleston from her home in New York City and to purchase "black clothes." The money was sent. However, over the course of several months, Mrs. King consulted six

4.12 *Aftermath of the collapse of one of the pneumatic caissons, December 2, 1929. Courtesy of the Charleston County Library*

*4.13 During construction the edge of the caisson hit a pocket of firm material and tilted 29 degrees out of vertical. Tons of mud and water rushed into the working chamber at the base of the caisson, killing seven bridge workers or "sandhogs." It took approximately two weeks to right the caisson. All of the bodies were recovered.* Courtesy of the Charleston County Library

separate New York attorneys in an effort to expedite the case. With each of Mrs. King's new attorneys, a frustrated John Cosgrove was forced to begin new lines of communication. The case dragged on, until Cosgrove, exasperated by Mrs. King's lawyer-swapping, settled with the Foundation Company for $3,600, considerably less than the original $25,000 requested. Cosgrove sent Mrs. King her portion of the money, less his fees and the $100 loan, which amounted to $1,400 (equivalent to $14,000 in year-2000 dollars). King's minor child received a portion of this money; Mrs. King received only $720.[19]

Fourteen men died during construction of the Cooper River Bridge: in addition to the seven sandhogs killed in the caisson accident, another sandhog died as a result of the bends, one worker was electrocuted, a steel worker fell from the high trusses, and four laborers died in other, unspecified construction accidents. The project did employ basic safety procedures, though they pale in comparison to what is required today. Photographs of the project indicate that bridge workers did not always wear hard hats and were not always using safety harnesses when working up on the 18-inch-wide steel beams of the trusses. Bridge workers did practice lifesaving skills by staging

*4.14  Construction of the Cooper River Bridge, May 29, 1929.* Courtesy of the Charleston County Library

*4.15  Steel workers balance on 18-inch-wide beams used in the trusses of the Cooper River Bridge.* Courtesy of the Charleston County Library

*4.16 Training dummies, like the ones shown in this June 8, 1929, photograph, were used in safety*
*exercises during construction of the Cooper River Bridge. The dummies were thrown off of the top*
*of the bridge and into the water; rescue crews then rushed to the dummies to practice lifesaving skills.*
Courtesy of the Charleston County Library

mock rescues of training dummies thrown from the bridge. Though the death of fourteen workers was tragic, the loss of life was not uncommon for major bridge construction at the time. During the 1907 construction of the Quebec Bridge a truss collapsed, killing 74 men. In 1917, the same bridge, then under reconstruction, lost another eleven workers when another truss collapsed. During construction of the Golden Gate Bridge, completed in 1937, considerable safety features were employed in an orchestrated effort to minimize accidents, yet eleven men died while working on that project.

## Construction of the Superstructure

Upon completion of the piers, work began on the trusses and road deck, collectively known as the superstructure. The Virginia Bridge and Iron Company supplied the 12,000 tons of steel used to construct the superstructure. A small train track was laid across the bridge deck to carry the material to the point of construction, where cranes lifted the steel beams into position. The cantilever spans were designed to be free-standing throughout erection. Photographs taken during construction show a massive crane perched on the edge of a freestanding but incomplete cantilever arm, seemingly in defiance of gravity. The approaches and the anchor arms were erected with the

*4.17 Town Creek span of the Cooper River Bridge, February 1, 1929, with work begun on the causeway over Drum Island (at left).* Courtesy of the Charleston County Library

*4.18 Construction of the anchor arm of the Town Creek span closest to the Charleston shoreline, December 28, 1928. A large crane, used to lift steel into position, sits at the end of the incomplete span. Beneath the crane temporary scaffolding, known as falsework, is used to support the anchor arm. Workers reached the construction site by climbing a staircase, shown at left.* Courtesy of the Charleston County Library

*4.19 During construction of the Cooper River Bridge, massive cranes were used to lift heavy steel beams into position. The cranes, which weighed more than 148 tons each, remain the single heaviest objects ever to cross the bridge. This photograph, taken on November 15, 1928, shows one crane as it inches its way along the incomplete Town Creek span.* Courtesy of the Charleston County Library

aid of temporary scaffolding, known as falsework. The wood-and-steel falsework supported the bridge sections until they could be made self-supporting. The entire process moved quickly, with crews working on the Town Creek and Cooper River spans nearly simultaneously.

During construction between 500 and 600 men worked on the bridge at any given time, working up to ten hours a day from Monday through Friday and eight hours on Saturday and earning, on average, $1.10 an hour (roughly $11 in year-2000 dollars). Many workers were from out of town and rented rooms in local residences. Though pay was good for Charleston in the 1920s, the work was hard. Longtime Charleston mayor Joseph P. Riley related his own family memory: "My father's youngest brother wanted to work on the bridge and finally got the chance. He lasted one day and quit and no one could blame him."[20]

4.20 Large cranes sit on the ends of each cantilever arm of the Cooper River span in this June 18, 1929 photograph. A four-masted sailing ship glides underneath the incomplete bridge. To many Charlestonians the Cooper River Bridge was the symbol of the city's emergence from the past into a modern, more progressive era. Courtesy of the Charleston County Library

*4.21 Progress on the Cooper River span, May 10, 1929. Work began on both shores of the Cooper River and proceeded nearly simultaneously. The bridge was joined in the center of the shipping channel.* Courtesy of the Charleston County Library

*4.22 Temporary scaffolding, or falsework, supports an incomplete anchor arm and crane in this May 17, 1929, photograph.* Courtesy of the Charleston County Library

*4.23 Workers lay one of the bearing shoes used to tie the bridge superstructure to the piers on May 17, 1929.* Courtesy of the Charleston County Library

## The Completed Bridge

Construction of the Town Creek span was completed on March 29, 1929; meanwhile construction was well under way on the larger span over the Cooper River. The Charleston papers, the *Evening Post* and the *News and Courier,* frequently ran stories with detailed technical reports on the progress of the bridge. The public thrived on every aspect of the construction. The Cooper River span was joined two months later, on June 29, connecting the entire bridge for the first time. Thomas Stevenson, a graduate of The Citadel and an engineer with the McClintic-Marshall Company, was the first person to walk across the completed structure. The final phase of the project was the installation of the reinforced concrete roadbed, which measures twenty feet wide (slightly wider at the curves) and is located between twelve-inch-high concrete curbs. The concrete for the road deck was supplied by concrete-mixing plants located at each end of the bridge. Railroad cars hauled the concrete to the work site. The paving took nearly a year to complete, commencing on July 27, 1928, and finishing July 22, 1929. The original barriers provided on the bridge still remain: three and a half-foot-high, horizontal side-rails. Lights were placed at 200-foot intervals on each side of the roadway. Lights were also placed on top of the bridge as a warning to low-flying aircraft.

On the Charleston peninsula the original approach to the Cooper River Bridge was located at the old intersection of Lee and America Streets. The bridge access leads to

4.24  *Completion of the concrete road deck on the Cooper River Bridge on July 17, 1929. Laying the concrete deck was the longest phase of construction, requiring nearly a year to complete. The road deck of the Cooper River Bridge is only 20 feet wide with two lanes for traffic; modern highway standards require automobile lanes to be at least 12 feet wide.* Courtesy of the Charleston County Library

*4.25 The Charleston approach to the Cooper River Bridge was originally located at the old intersection of Lee and America Streets. This photograph, taken April 28, 1929, shows how the intersection looked at the time the bridge was being constructed.* Courtesy of the Charleston County Library

a 457-foot long concrete trestle, or elevated roadway. From this point a steel trestle extends 1,628 feet until it reaches pier number one, which is one of the two anchor piers of the Town Creek span. The rising grade at this point is 6 percent (6 feet vertical rise to each 100 feet horizontally) to a height of 135 feet. From the center of the Town Creek span the downgrade is 5 percent to the center of Drum Island, at which point the bridge deck is only 50 feet above ground level. On the eastern shore of the Cooper River, the bridge originally tied into Route 40 in Mount Pleasant, which was eventually renamed Coleman Boulevard. To construct the Mount Pleasant approach contractors excavated nearly three-quarters of a mile of marsh to a depth of ten feet, removing nearly 59,000 cubic yards of soft mud and replacing it with firmer material. From the Mt. Pleasant shore a steel trestle 1,342 feet long rises at a 5 percent grade to pier thirteen. Here three deck spans (simple trusses underneath the road deck) join and extend to pier ten, one of the anchor piers for the Cooper River span and the pier that collapsed during construction. The other anchor pier is pier seven, while piers eight and nine are massive cantilever piers with foundations nearly 100 feet below the water surface.

Perhaps the most unusual feature of the Cooper River Bridge is its thirty-four degree turn over Drum Island. To minimize the erosive effects of river currents, which could undermine the pier foundations, the bridge piers were installed perpendicular to the flow of the river. The turn over Drum Island allows the bridge to cross both

Town Creek and the Cooper River at a perpendicular angle. The infamous roller-coaster effect is created by the steep drops to the Drum Island viaduct. By running the viaduct just 50 feet above the ground of the uninhabited island, Hardesty was able to reduce material needs, and thus the overall cost, of the bridge. Upon completion engineer Charles K. Allen remarked, "It's our first roller-coaster bridge." The name stuck, and the bridge has since been known by the colorful nickname "The Old Roller Coaster."[21] The lanes of the Cooper River Bridge are only ten feet wide, designed to carry two-way traffic. By comparison, modern highway standards require all new traffic lanes to be at least twelve feet wide. Today bridge designs are limited to a maximum 5 percent grade. The steep dips, curve, and narrow lanes make driving the Cooper River Bridge similar to riding a roller coaster—with the added thrill of being able to easily view the water far below.

Upon completion, the Cooper River span was the third largest cantilevered-truss bridge in the world serving automobile traffic (the fifth largest of all railroad and automobile cantilevered-truss bridges), exceeded only by 1,182-foot-long Queensboro Bridge in New York and the 1,100-foot-long Carquinez Strait Bridge near San Francisco. The Quebec Bridge and Scotland's Firth of Forth Bridge remain the two

*4.26 The incomplete bridge as it appeared from the top of the Cooper River span on March 27, 1929. The Town Creek span and Drum Island causeway can be seen in the background. The photograph shows the curious contour of the bridge. The undulating grades have caused the Cooper River Bridge to be nicknamed "Old Roller Coaster." The 35-degree curve over Drum Island allows the bridge to cross Town Creek and the Cooper River at a perpendicular, thereby minimizing the erosive effects of river currents on the bridge's piers.* Courtesy of the Charleston County Library

| | |
|---|---|
| Opened | August 8, 1929 |
| Bridge design | Double Cantilever |
| Construction time | 17 months |
| Surveys begun | February 7, 1928 |
| Groundbreaking | May 19, 1928 |
| Owned by | The Cooper River Bridge Corporation |
| Designed by | Waddell and Hardesty, New York, N.Y. |
| Cost | $5.7 million |
| Length | 2.7 miles |
| Materials used in construction | 26,562,000 pounds of steel<br>52,797 tons of concrete |
| Vertical clearances | 150 feet above the Cooper River<br>135 feet above Town Creek |
| Horizontal clearances | 1000 feet over the Cooper River<br>600 feet over Town Creek |
| Roadway | Two lanes, each 10 feet wide. Roadway has 12-inch-high concrete curbs and 3.5-foot-high side rails |
| Bought by Charleston County | September 1, 1941<br>Cost: $4.4 million |
| Bought by the State of South Carolina | March 8, 1945<br>Cost: $4.15 million |

*"My first trip across the bridge, I was afraid because it is so narrow. My boyfriend at the time knew that I was nervous so he leaned over and kissed me to keep me from thinking about it (we were passengers). Now every time I cross the bridge I think back fondly to him."*

—Beth Kistelnik, student, College of Charleston, 1996

*4.28 Aerial photograph taken on May 9, 1929, showing progress on the Cooper River Bridge. The Town Creek span (in foreground) is nearly completed, while work continues on the Cooper River span. The city of Charleston is seen at right.* Courtesy of the Charleston County Library

longest cantilevered truss bridges. At the time, the Town Creek span alone was the twelfth longest in the world. The construction of the Cooper River Bridge attracted national attention from newspapers and magazines. Paramount Studios even came to Charleston to film a silent newsreel about the construction of the bridge. Charlestonians thrived on the excitement. One local publication proudly stated that the new bridge was "fifteen feet higher than the Brooklyn Bridge. The largest ship afloat can pass under this bridge."[22] The *News and Courier* boasted of the length of the structure, right down to the fractions: 14,374.2 feet long, or 2.7224 miles. Construction superintendent Charles K. Allen declared the bridge's strength by stating that "if a 10,000 ton steamer should collide with the . . . bridge it would be the steamer alone which would suffer damage," words that would be dramatically disproved nearly twenty years later.[23] The most impressive figure of all was the length of time required to build the bridge. Construction was completed under budget and in record time, just seventeen months after the first survey and three months ahead of schedule. After decades of waiting, Charleston had a direct land route to the north. On August 8, 1929, opening day for the Great Cooper River Bridge, Charleston was ready to celebrate.

## CHAPTER FIVE

# HOPE AND DESPAIR

*"As children, we tried to hold our breath or our feet up all the way across."*

—Dorinda Harmon, James Island, 1996

## The "Great Cooper River Bridge" Is Opened

On Thursday, August 8, 1929, the first automobiles crossed the Cooper River Bridge, almost three months ahead of the projected completion date of November 4, 1929. Charleston loves a party, and a grand, three-day gala was planned for the opening of the bridge. In a seventy-page supplement to the *News and Courier*, the paper and its advertisers proclaimed their pride. "The Great Cooper River Bridge!" boldly stated one headline. One advertisement summed the atmosphere: "[Charleston] is now taking her place among cities."[1] The bridge was the city's gateway to the future.

Between 25,000 and 30,000 people from across the state were on hand for the opening festivities. The city of Charleston footed the bill for the celebration and newspaper headlines announced, "Visitors Will Be Greeted and Treated with Conventional Hospitality."[2] Banks, city offices, and businesses closed for the long weekend. The opening ceremonies were planned and executed by twenty-two separate committees, including the Committee on Motion Pictures and Press, the Committee on Airplanes and Air Maneuvers, and the Committee on Auto and Motor Boat Races and Sparring Exhibitions. The official ribbon-cutting ceremony was held on August 8, followed by an exhaustive program of events designed to showcase modern Charleston. After an opening military parade and a two-and-a-half-mile-long naval parade, a twenty-five mile automobile race was held on the Isle of Palms. Later, a motorboat race was held in Charleston Harbor. John Grace, the president of the Cooper River Bridge Corporation, wore his trademark white summer suit and black bow tie while serving as master of ceremonies. Grace compared the opening of the Cooper River Bridge to Columbus's discovery of America, proclaiming, "It is the mission of civilized man to conquer nature's barriers. There is a 'pleasure in the pathless wood' . . . but these days we are soon aroused from such poetic reveries by perhaps the static hum of the radio or the hum of some . . . aeroplane."[3]

Local officials, state dignitaries, and celebrities praised Charleston's newest accomplishment. Those on hand included Charleston mayor Thomas Stoney; Governor

*5.1 Brochure for the opening of the Cooper River Bridge, August 8–10, 1929. The opening of the bridge was celebrated for three days. Festivities included automobile, motorboat, and airplane races as well as parades and dances.* Collection of Pamela Gabriel

John G. Richards; U.S. Senator Coleman L. Blease; State Highway Commission chairman C. E. Jones; and two state senators, A. G. Butler and William S. Legare. Colonel James Armstrong, a former Confederate officer, balanced on crutches and one leg to cut the ribbon prior to the first car's crossing the bridge. The lead car carried Grace and his old antagonist, Mayor Stoney. A musical program opened the parade, and another concert was held at the Isle of Palms Pavilion to conclude the day's events. The bridge opening was captured oń film by Paramount Studios and shown in newsreels to audiences around the world.

The opening celebration was a resounding success and continued on Friday, August 9. The second day of the Cooper River Bridge Celebration began with the Historic Floats Parade. Participants dressed in period costumes and traveled down King Street to the Battery, up East Bay Street and over Broad Street, where the review stand was set up. Then it was back up Meeting Street to Lee Street and over the bridge to Mount Pleasant. The parade was designed to illustrate the ease with which motorists could now travel the Coastal Highway (Route 40) into Charleston via the Cooper River Bridge. After a stop in Mount Pleasant the parade continued on to McClellanville, Georgetown, and ended in Conway—a total route of over 100 miles. In the afternoon there were more motorboat races at the Isle of Palms, another musical presentation and a sparring exhibition featuring three events. The first matched high school contenders, the second featured military opponents, and for the third event, a college student was pitted against a soldier from Fort Moultrie. The owners of the bridge, who also owned the Isle of Palms, wanted to showcase their resort, and

5.2 *Corporate logo of the Cooper River Bridge, Inc.* Collection of Pamela Gabriel

each day's events were scheduled to end on the island. In addition to boat and car races held during the day, music and dancing was held at the Isle of Palms Pavilion each evening of opening weekend and lasted until two in the morning.

On the final day of the celebration, Saturday, August 10, Charleston pulled out all the stops and put on a show designed to dazzle as the city marked its arrival into the twentieth century. Timed to coincide with the bridge opening, the Charleston Airport was dedicated as part of the bridge celebration. Located just ten miles outside the city, the airfield consisted of a 3,000-by-3,500-foot grass-surface flying field marked by a 100-foot circle of wind cones. Six events featured stunt flying, spot landing, races, and a demonstration by Army, Navy, and Marine Corps planes. Late in the day the celebration continued at the Isle of Palms with a lifesaving exhibition, more motorboat races, and more musical programs, concluding with a spectacular fireworks display.

The late Jack Leland, a journalist for the Charleston *Post and Courier*, recounted his eyewitness account of the bridge celebration in an article published in 1989. Thirteen years old at the time of the bridge's construction, Leland recalled the fascination of seeing "the caissons rising from the water" and then attending the opening festivities. His remembrance offers a delightful tale of the day:

There was no question we would be there. Ours was an open "touring car," or phaeton, with a waterproof canvas top that could be lowered, isinglass curtains, wide running boards and hand crank. It was roomy and slow and dependable. That day, there were picnic baskets filled with such goodies as pickled shrimp, deviled crabs, fried chicken, sandwiches, salad and cake.

Tied to the back was a large, galvanized wash tub that was filled with ice from the ice house in Mount Pleasant at the southwest corner of Pitt Street and Rifle Range Road (today's McCants Drive) . . . it contained a watermelon, jars of tea, lemonade, soft drinks that my father referred to as "belly wash" and a number of bottles of "home brew," the potent and illicit beer citizens made to offset the Saharan dustiness of Prohibition.

The ice tub complete, we drove past the "old car barn" that had served the trolley line that once went to Sullivan's Island and Isle of Palms, through the Lucasville portion of Mount Pleasant, across Shem Creek on a new wooden bridge, through Peach Orchard Farm (Baytree and The Groves), past the site of the ancient tavern at Mathis Ferry and onto the newly paved road leading to the east end of the bridge. The great bridge towered up in the sky ahead of us and we settled in with the other East Cooperites to await the grand procession out of Charleston. I remember a military band and the long line of cars carrying the dignitaries, including Mayor Thomas Stoney . . . Cap'n Tom, as we knew him, was an old family friend and we really whooped it up as he rode by.

John P. Grace, a former mayor and bridge company official, was with the "high mucky-mucks," as we called them. . . . He and Stoney were bitter enemies. I recall my father telling someone that he would bet it was the first time they rode in the same car together. . . . Finally, the time came for the East Cooperites to cross the bridge to Charleston. The bridge . . . was a great event for people living east of Charleston.

In his account, Leland quotes the impressions of former longtime Santee resident and South Carolina poet laureate Archibald Hamilton Rutledge: "I was born in the week-and-month travel period and have lived in the day-and-week travel period. Now we are in the hour-and-minute travel era."[4]

On opening day approximately 11,000 automobiles crossed the bridge during the initial four-hour toll-free period. In the first hour and forty-five minutes after tolls were in effect, another 4,000 cars crossed the structure. The tolls for the bridge were in line with the fees charged by the Cooper River ferries: fifty cents per vehicle and driver, and fifteen cents for each additional passenger.[5] Children under the age of six were free. Empty trucks were charged seventy-five cents, and when loaded the cost was $1.50. Motorcycles had a bargain rate, a mere thirty-five cents. The tollhouse was located on the Mount Pleasant side of the bridge and collected fares for vehicles going

*5.3 Charleston Mayor Thomas Stoney (left) watches as John P. Grace (center) and Confederate Col. James Armstrong (right) cut the ribbon to open the Cooper River Bridge on August 8, 1929.* Courtesy of the Charleston County Library

*5.4 Photograph taken on July 16, 1929, showing Paramount Studios filming the new Cooper River Bridge shortly before the bridge's official opening. The construction of the bridge attracted international attention, and Paramount filmed it in various stages of completion for use in newsreels. An acquaintance wrote John Grace to report that he had seen newsreel coverage of the bridge's opening ceremonies while vacationing in Bremen, Germany.* Courtesy of the Charleston County Library

*5.5 Historic Floats Parade for the opening of the Cooper River Bridge, August 9, 1929.* Courtesy of Sandy Knisley, from the collection of Robert Legaré Coleman

*5.6 The Historic Floats Parade started in Charleston and continued for over one hundred miles, ending in Conway, S.C. Stops were made in Mount Pleasant, McClellanville, and Georgetown. In the sweltering heat some of the parade vehicles overheated when crossing the bridge's steep grades.* Courtesy of the Charleston County Library

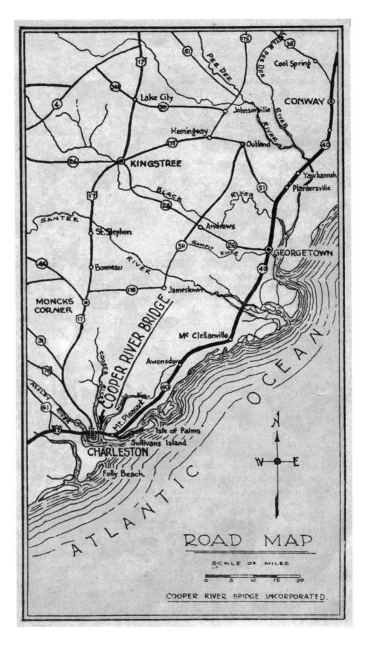

5.7 *On opening day, bridge owners and Charleston city officials emphasized the fact that the Cooper River Bridge was an important link in the Coastal Highway (called Route 40 in South Carolina). This map, published for opening day, August 8, 1929, is one of several that highlighted the bridge's importance.* Collection of Pamela Gabriel

both to and from Charleston. The speed limit was 20 mph (later increased to the breakneck speed of 35 mph). Passing was prohibited, as were pedestrians and bicycles. Horses and horse-drawn wagons were allowed on the bridge only by permission of the bridge administrators.

*5.8 Widespread excitement surrounded the opening of the Cooper River Bridge. So momentous was the occasion that it inspired local publisher Furman C. Moseley to write an "ode" to the bridge in September 1929.*

High o'er the Cooper stands a monster of steel.

To look at its greatness compels us to feel

That surely no mere human hand could conceive

What stands before our eyes, but is hard to believe.


'Tis the labor of men whose dreams have come true.

They frantically worked many a night thru.

Altho it was not by dreams that we see

This huge superstructure that leads to the sea.


Perseverance and skill were required of the men

Who planned and worked for the glorious end.

And our souls breathe a prayer for the brave men who died

That our children might over the Cooper River ride.


There it stands, tow'ring high o'er the waters of the bay

Hurling a challenge to those who would say

"It shall never be done, the cost is too great."

And now they ride over on steel and concrete!


May the far-sighted men who labored and dreamed

Of this wonderful structure across this stream,

Therein see a monument of which to be proud—

Casting forever its head in the clouds!


May the sacrificial men who gave of their lives,

Who left their children, their homes and their wives,

Forever find peace with the Maker of Men

In the land of the blest—where we live without sin.


The Cooper River Bridge—oh, long may it stand—

The highest and longest of its kind in the land;

May her beautiful peaks ever each to the sky,

That the pessimist may marvel as he rides by!

5.9 *Postcard, circa 1940, showing the Charleston Airport. The opening of the airport was timed to coincide with that of the Cooper River Bridge, on August 10, 1929. The airport grew little throughout the 1930s.* Collection of Pamela Gabriel

5.10 *A badge worn by spectators attending the opening ceremonies for the Cooper River Bridge, August 8–10, 1929. The bridge tied into Route 40 in Mount Pleasant, known as both the Coastal Highway and Washington Highway.* Photograph by Lindsay Richardson and courtesy of Sharon Hite, John's Island, S.C.

5.11 *Automobiles cross the Cooper River Bridge during the four-hour, toll-free period on opening day, August 8, 1929. Note that cars are crossing the bridge in both directions. Until 1966, when a second bridge was opened, the Cooper River Bridge served two-way traffic. Courtesy of the Charleston County Library*

5.12 *The Cooper River Bridge's tollhouse, shortly after the bridge was opened. The tollhouse was located on the Mount Pleasant side of the bridge. Fares were collected from cars traveling in either direction. When the bridge opened in 1929 the fare was fifty cents per vehicle and driver and fifteen cents for each additional passenger. Many locals recall hiding in vehicle trunks to avoid the additional passenger fees.* Courtesy of the Charleston County Library

## The Hopes of the '20s, the Reality of the '30s

The bridge promoters envisioned throngs of visitors crossing the Cooper River Bridge to enjoy the delights of their beach resort at the Isle of Palms. Others had their own visions. Local businesses touted the event with oversized advertisements in the *News and Courier* proclaiming their hopes and dreams:

> The construction of the bridge will be of tremendous advantage and benefit to the entire northeastern portion of the state . . .

> The Cooper River Bridge Marks a New Era in Charleston's Progress . . .

> New Business, New Industries, New Visitors . . . this giant structure (one of the largest of its kind in the world) not only opens the ocean highway linking Charleston to the North and South by land as well as by water, but also makes even more practical the great industrial development of the east bank of the Cooper River and the territory beyond.[6]

The residents of the village of Mount Pleasant were excited about the possibilities the new bridge offered. At the opening celebration Mount Pleasant mayor T. G. McCants expressed the optimism of the town when he declared that the bridge would "enlarge our world" and offer boundless opportunities. He declared the desires of

Mount Pleasant for "new people and new homes" but emphatically opposed "new industry" in his town. Residents of Mount Pleasant looked at the new development of West Ashley and imagined the possibilities for East Cooper. By 1929, just three years after the Ashley River Bridge opened, the new neighborhoods of West Ashley had seen significant growth: Riverland Terrace had a population of approximately 250 and Windermere's population was 175. There were other new subdivisions as well: Edgewater Park, Wappoo Heights, Wappoo Hall, Woodland Shores, and Crescent Heights. West Ashley also had a new country club and municipal golf course.

Excitement surrounding the bridge's opening spread throughout East Cooper. H. T. Morrison, the mayor of McClellanville, a tiny hamlet north of Mount Pleasant, said at the bridge's opening, "There is no place of like size to which the Cooper River Bridge means as much as it does to McClellanville . . . to McClellanville it is the outlet to the world. Without it we are an isolated community. With it, we become a part of the world."[7] On January 1, 1930, despite the crash of the stock market only months before, the *News and Courier* predicted that in the summer of 1930 "the influence of the Cooper River Bridge will be felt for the first time, [and] Charleston's beach resorts . . . will attract many more visitors than heretofore."[8]

Cast in the light of the bustling economy in the 1920s, the success of the Cooper River Bridge both as a business venture and in promoting Charleston's growth and development seemed assured. No one anticipated the lasting effects of the Great Depression. Perhaps John Grace possessed the Midas touch in reverse: his Florida land scheme had met with financial disaster, and his newspaper had failed. Now, less than three months after the completion of the Cooper River Bridge, the entire country was struck by economic calamity. Despite the bold predictions by the *News and Courier*, there would be no flood of wealthy tourists heading south. East Cooper experienced no great development, and no prosperous investors rushed to purchase land on the Isle of Palms. No manufacturing plants were relocating to the Charleston area; no throngs of vehicles were crossing the Cooper River Bridge. Sufficient toll revenues never materialized, and the hopes for big bridge profits for Grace and the bridge financiers crashed along with Wall Street.

Within only six weeks of its opening, the Cooper River Bridge was at risk of being auctioned for delinquent county taxes, on September 20. Thanks to the support from Midwestern financiers, who were majority shareholders in the bridge, the Cooper River Bridge Corporation managed to pay the overdue taxes and save its investment, but the bridge's financial health only grew worse. Between 1929 and 1932, the bridge's yearly toll revenues dropped from $231,000 to $188,000. Yearly interest payments due bondholders amounted to nearly $220,000, and maintenance and operations, averaging $70,000 per year, weighed heavily on the bridge owners. At the end of 1932, the Cooper River Bridge Corporation was $239,000 in debt and could not afford interest

*5.13 As Charleston struggled through the Depression, numerous tourism guides were published in an effort to lure visitors to the city. Maps such as the one shown above illustrated how Charleston's bridges allowed motorists easy access to the city center. Tourists were not always happy about Charleston and its bridge; northern visitor William Oliver Stevens commented critically in his travelogue* Charleston *(New York: Dodd, Mead and Company, 1939): "Having . . . already [paid] twenty-two cents a gallon for our fuel in South Carolina, we were not surprised to find that there is a toll for the privilege of using the bridge; to wit, sixty-five cents for a car and two passengers." Stevens also criticized Charleston's high prices. "The rates? Breakfast, one dollar; lunch, two dollars; and dinner, three dollars. By the day the rate for two people, rooms and bath, is forty dollars. Forty dollars a day!" Editorializing about Charleston's few affordable hotels, Stevens lamented, "There is a bathroom—as there is a Happy Land—far, far away . . . if atmosphere is what you want most in Charleston you will be happy, even though uncomfortable."* Collection of Pamela Gabriel

payments to its bondholders. In 1934, bondholders agreed to temporarily readjust the interest rate on the company's bonds from 6 percent to 2 percent. Restructuring effectively eliminated dividend payments to the bridge's stockholders.[9] This reduced much of the debt owed by the Cooper River Bridge Company, but the bridge continued to struggle. By 1935 the company was almost $360,000 in debt. In that year, the new president of the Cooper River Bridge Corporation, William Pohl, prevailed upon the city and county of Charleston to grant the bridge a ten-year tax moratorium. Increasing tolls was never an option, as state law had locked them in place. In 1936, after attempts by bondholders to place the bridge into receivership, the Cooper River Bridge Corporation, then $400,000 in debt, declared bankruptcy. Stock in the Cooper River Bridge Corporation that originally sold for $50 per unit was now worthless. Bonds that had originally sold for $1000 now traded at $160.

The complete economic failure of the bridge as an investment came as a surprise to both the bridge owners and financiers. During the planning for the bridge, financiers conducted independent traffic assessments and concluded that the Cooper River Bridge would be profitable. The bridge owners gave only passing consideration to a full traffic analysis, but in all likelihood both the owners and financiers based their assessments on passenger volumes carried by the Cooper River ferries. The bridge promoters were certainly aware that in 1927, the year that plans for the Cooper River Bridge were being finalized, the Cooper River ferries were carrying more than 526,000 persons and more than 100,000 automobiles, earning a record profit of nearly $25,000.[10] As recorded in the annual statements of the Cooper River Bridge Corporation, the bridge owners realized that the success of their project likewise depended on summer tourism. Reports from the Cooper River Ferry Commission, the quasi-governmental agency that administered the ferries, indicated that demand for ferry service surged in the summer months. In 1925, monthly vehicular traffic on the ferries ranged from a low of 1,294 in February to a high of 6,859 in August. Over half of the Ferry Commission's 1927 revenues were earned in the summer months, when the ferries were carrying a peak of nearly 13,000 automobiles each month.

Though the Depression curtailed the flow of tourism to the beach and was a central factor in the bridge's economic collapse, data from the Cooper River ferries indicate that even without an economic depression, there was not enough traffic to meet the bridge's debts. The yearly liabilities for the Cooper River Bridge were enormous, over $300,000 per year for operations and debt repayment. By comparison, the Cooper River Ferry Commission operated with relatively low expenses; the total operating expenses for the commission in 1927, including debt repayment and maintenance costs, was roughly $115,000. The Cooper River Ferry Commission received roughly $139,000 in toll revenues in 1927.[11] Just to meet its costs, the bridge company needed revenues equal to twice the yearly income generated by the ferries in 1927.

*5.14 The Cooper River ferry terminus in the 1930s. Cross-harbor ferries operated until 1939 and competed with the bridge for toll revenues.* Courtesy of the Avery Research Center, College of Charleston

Bridge owners counted on out-of-town motorists to boost toll receipts, but with the onset of the Depression the motorists stayed home. Nor could the bridge count on local traffic to meet its revenue objectives. With fewer than 1,600 residents and even fewer commuters, Mount Pleasant provided a small customer base. Even the truck farmers east of the Cooper River avoided the bridge and the tolls that ate into their profits, opting instead to take the long, but toll-free, inland route to Charleston.

Other reasons help explain the economic failure of the Cooper River Bridge. The Cooper River Ferry Commission's network of ferries allowed families to travel to the beach without requiring ownership or rental of an automobile. In 1925 there were only 6,400 automobiles registered in Charleston County, and though automobile ownership in Charleston rose throughout the 1920s, it remained out of reach for families struggling through the Depression. Traveling as pedestrians, families could ride the ferries for only ten cents per person, as compared to the bridge toll of fifty cents per vehicle and fifteen cents per extra passenger. In 1929 an agreement was reached that allowed the Cooper River Ferry Commission to continue to operate after the Cooper River Bridge was opened. Vehicle tolls on the ferries were not permitted to be lower than tolls charged on the bridge. At the time, the director of the Cooper River

Ferry Commission, Alfred Halsey, pleaded the case for pedestrians who relied on the ferry services. Halsey also noted that older automobiles and heavily loaded trucks could not navigate the steep grades of the Cooper River Bridge, and were therefore reliant on ferry service. The Cooper River ferry terminals were also located close to Charleston's city center, and the Cooper River Bridge was located outside the city proper. Visitors often opted for the more direct route to the beaches via the harbor ferries. The Cooper River Ferry Commission closed its operations in 1930, but other ferries, such as the *Nansemond*, continued to operate until the end of 1939, directly competing with the bridge for toll revenues.

## Purchase and Repurchase

In the early 1930s state transportation officials began to pursue the idea of buying the Cooper River Bridge.[12] State officials recognized the hopeless financial condition of the bridge and floated the notion that the structure could be purchased for about $1 million, based on the low prices of the bridge's bonds. They were supported in their efforts by Charleston's civic and business leaders, who believed that the bridge tolls were choking the city's prosperity and limiting the development of new roads south of the city. Charleston Mayor Burnet Maybank wrote in 1933, "it is useless to attempt any . . . [new] road as long as the bridge is privately owned and a high toll is charged. . . . I certainly would not care to spend a great deal of the State's money to enhance the value of the Cooper River Bridge for the bondholders."[13] In 1934, lowcountry legislators introduced a bill authorizing the state to buy the bridge, but Governor Olin Johnston and his upstate South Carolina allies fought the legislation. They argued that the privately owned Cooper River Bridge was not the concern of the state and that purchase of the bridge would give Charleston a disproportionate amount of precious state highway funds. Opponents labeled the bridge a "white elephant" and a "chamber of commerce project" intended to help businessmen and lawyers at the expense of poor farmers.[14] Said one upstate representative of the idea to purchase the bridge, "We can't afford to pamper one small section of the South Carolina . . . at the expense of the rest of the state."[15] Lowcountry politicians countered that sound management of the bridge was an economic necessity for the Charleston area and would help the state highway department better manage regional highways. For the next ten years the fate of the Cooper River Bridge became a political football, pitting upstate legislators against lowcountry politicians. The General Assembly considered bridge purchase bills at least four times and defeated them all. The debate was another chapter in the generations-old regional rivalry between the lowcountry and upstate, a characteristic of South Carolina politics.

As the bridge purchase debate dragged on, the price of the bridge increased. By 1938, the proposed cost of the bridge had more than doubled from its 1934 price to $2.5 million. In 1941, state legislation was passed allowing Charleston County to purchase the bridge. The county moved quickly and paid $4,400,000 for the structure. Bondholders, many of whom managed to make respectable profits from their original investments, greeted the sale with enthusiasm. Bondholders were paid $1,237.50 on an original $1,000 bond. Many investors had purchased bonds on the open market for as little as $160. Stockholders fared much worse. The original price of a unit (one share preferred, one share common) of stock in the Cooper River Bridge Corporation was $50. After the purchase by Charleston County, stockholders were paid $5.95 per unit.

The reason for the steady increase in selling price, from $1 million in 1934 to $4.4 million in 1941, was due partly to the strengthening of the bridge's balance sheet. After bankruptcy and debt restructuring cut much of its debt, the bridge was able to meet its financial obligations, though it never made significant profits. Throughout the late 1930s, as the Depression waned and the economy slowly strengthened, toll revenues began to increase. In 1933 gross revenues amounted to only $146,000. By 1937 revenues had surged to $249,000, and by 1940 they peaked at $299,000. During the period between 1920 and 1940 the permanent population living east of the Cooper River remained constant, indicating that the rise in bridge toll receipts could not be explained by an increase in the number of commuters traveling to and from Charleston. Bridge administrators attributed increased earnings to a series of mild polio seasons and a corresponding increase in summer beach traffic.[16] Likewise, earnings on bridge bonds began to increase, from a low of 2.46% in 1933 to 6.39% in 1940. In the bond market, prices for bridge bonds rose, reflecting the improved fiscal health of the bridge enterprise.

Political forces also helped to boost the selling price of the Cooper River Bridge. Throughout the 1930s a strong anti-toll sentiment fermented in Charleston, and area business leaders grew to believe that the bridge tolls were stifling economic development and tourism. In 1933 Mayor Maybank declared, "The entire future of Charleston absolutely depends upon the Cooper River Bridge being made free . . . if we are to survive as a tourist town this bridge must be free."[17] As Charlestonians increased political pressure on their leaders to purchase the bridge and, most importantly, remove the tolls, the bridge owners were in a more comfortable bargaining position than they had been in the early 1930s.

Implicit in the county's decision to purchase the bridge in 1941 was the expectation that the state would immediately move to buy the structure. In 1940, a legislative proposal attempted to guarantee state purchase of the bridge after the county assumed ownership. Again, upstate legislators blocked the sale, and the legislation died at the end of the session. In 1942, new legislation was introduced that would have allowed the state to purchase the bridge from the county. Again, the proposal was defeated.

Throughout World War II, the county continued to operate the bridge as a toll bridge, the only one in the state. Charlestonians continued to bemoan the onerous tolls. During county operation, from 1941 to 1945, yearly toll revenue remained at $300,000 per year, just covering operating and debt repayment costs.

John P. Grace, the former mayor and force behind the construction of the bridge, died on June 25, 1940, and did not live to see the great bridge-driven prosperity he forecasted in 1929. In 1943, by an act of the state legislature, the Cooper River Bridge was officially renamed the John Patrick Grace Memorial Bridge.

## The Bridge Is Free

When state senator Oliver T. Wallace of Charleston was reelected in 1943, he announced that his "lifetime ambition" was the removal of the bridge tolls.[18] By then the state had the hope of partial federal funding to assist in the bridge purchase. Federal funds were granted only to purchase toll-free bridges; to be eligible for the additional money, the state had to first purchase the bridge and remove the tolls. On March 8, 1945, despite unsuccessful attempts by upstate politicians to kill the legislation, a bill was approved by the General Assembly authorizing the highway department to purchase the bridge from the county. Charleston County was happy to unload the structure and sold it to the state for $4,150,000. The tolls were to be removed on July 1, 1946. The state formally assumed control of the John P. Grace Memorial Bridge on April 4, 1945.

Tolls on the Cooper River Bridge were officially lifted on a rainy June 29, 1946, and Charleston celebrated. Senator Wallace "happily" paid the last toll and Governor Ransome Williams declared the bridge "free of toll." The four sisters and brother of the late John Grace were special guests of honor for the ceremony. Six young girls from both Mount Pleasant and Charleston cut the ribbon on the new "toll-free" bridge. The scissors used to cut the ribbon were given to each of Grace's siblings and to Senator Wallace. Celebratory events included a wrestling match and a fish fry. A "toll-free" bridge parade was delayed but eventually held despite the rain. Parade participants came from as far away as Georgetown, S.C., to celebrate the "economic liberation" of the area between Charleston and their city. After the parade, drenched parade participants arrived in Mount Pleasant for the fish fry only to discover that the food had already been eaten by spectators unwilling to wait out the rain. Appropriately, the toll-free festivities ended on the Isle of Palms with a wrestling match, fireworks, and a grand ball.

Removal of the tolls marked the end of a significant chapter in the history of the Cooper River Bridge. Just as the Cooper River once created a natural barrier around Charleston, Charlestonians believed that tolls had created an economic barrier to the prosperity of their city. City officials and business leaders argued that the tolls

5.15 *Photographs from the toll-free celebration, June 29, 1946. The picture in the upper left shows state representative James Lofton of McClellanville, a longtime toll opponent, painting out the rate sign at the tollhouse. To the right, state senator Oliver Wallace pays the last toll on the bridge. The center picture shows dignitaries assembled for the ribbon-cutting ceremony to open the toll-free Cooper River Bridge. At the bottom, one of the floats that participated in the toll-free parade was from Georgetown, sixty miles north of Charleston, an indication of the widespread impact the removal of tolls had on the lowcountry.* Courtesy of the Charleston County Library

*5.16 The Cooper River Bridge as it appeared in the 1940s. The bridge was officially renamed the John P. Grace Memorial Bridge in 1943, in honor of the former Charleston mayor, a driving force behind the bridge's construction.* Courtesy of Greyscale Fine Photography Center, Charleston, S.C.

discouraged tourists and trade from coming into Charleston. Farmers east of the Cooper River complained that the tolls frustrated the growth of agriculture and related industry. Indeed, the area east of the Cooper River remained in isolation. Low-country residents termed the region South Carolina's "forgotten county" and blamed the much-cursed bridge tolls for hindering its development. In many respects, the Cooper River Bridge in the toll era was a failure. It failed to deliver profits for the bondholders and the stockholders in the Cooper River Bridge Corporation. It failed to usher in the runaway growth and development of the Charleston area as envisioned on opening day. Finally, the bridge even failed to deliver the great revival of the Isle of Palms.

The end of the bridge toll era signified the beginning of a more prosperous time for Charleston. By the time tolls were lifted in 1946, America had ended a world war and was entering a time of tremendous growth and economic expansion. The Cooper River Bridge would at last serve to spur the growth of Charleston to areas east of the Cooper River. It is ironic that the Cooper River Bridge in postwar era would not only facilitate great growth, but again would come to be viewed as a hindrance to Charleston's full development.

## CHAPTER SIX

# THE BRIDGE COMES OF AGE

Local leaders viewed the tolls on the John P. Grace Memorial (Grace) Bridge as the last hurdle to the potential growth of the East Cooper area. Shortly after the state purchased the bridge in early 1945, the *News and Courier* reported on development in Mount Pleasant, and predicted that after tolls were removed the East Cooper town would become the "second Windermere," referring to the burgeoning suburban area that developed west of the Ashley River at the foot of the Ashley River Memorial Bridge.[1] Even before the bridge was rendered toll-free, opportunists acquired tracts of land surrounding the village of Mount Pleasant. Real estate prices began to climb.

*6.1 Increased accessibility of areas north of the Peninsula brought not only population growth, but its darker side effects. Mount Pleasant's worries concerning "decadence" were not unfounded as vice crimes continued to thrive throughout the county. Marion J. Schwartz, director of Charleston County Police, destroys slot machines beneath the Cooper River Bridge in a circa 1959 crackdown on illegal gambling.*
Courtesy McKissick Museum, University of South Carolina

Though some residents expressed concern that the free bridge would allow unchecked access to "honky-tonks" and other alleged vices in Charleston, the small town embraced an anxious, optimistic mood. In the midst of this speculative fervor, tragedy befell the Grace Bridge. Mount Pleasant and the rest of Charleston experienced a painful lesson on their daily reliance upon the bridge.

## The *Nicaragua Victory* and Tragedy

On February 24, 1946, an overcast Sunday afternoon, just months before the tolls were to be lifted from the Grace Bridge, the 12,000-ton ship *Nicaragua Victory* drifted out of control during a sudden gale and slammed into the bridge. As the storm's strong winds blew, the ship drifted from its anchorage in the Cooper River. The ship's crew did not immediately realize their situation, but when they did the captain ordered an additional anchor to be dropped. It did little to stop the ship, which floated towards the bridge. The engines were off; the crew had no way to control the ship's movement. Soon the bow became stuck in the mud near the Mount Pleasant side of the Cooper River, and the stern pivoted around and slammed against the side of the bridge.[2]

One of the crewmen aboard the *Nicaragua Victory* described the collision the day after it occurred. "There was a thud, with little shock. Then there was a sickening, low groaning, grinding noise, which sharpened into a screeching tone. That was accompanied by the cracking and popping of the concrete. The big splash came at the same time with a flash of light when the high tension [electric] wires [that crossed the bridge] were snapped and then short-circuited. Then the lines hit the deck . . . there were sparks flying, and more fireworks."[3] Debris rained down upon the ship. The impact damaged one of the piers of the Cooper River span and carved out a 240-foot gap in the roadbed.

At the time of the collision, automobiles were crossing over the bridge. According to eyewitness accounts of the accident, when the ship struck the bridge, vehicles skidded to a halt. One car could not stop in time. Elmer Lawson, a quarterman and electrician at the naval base, his wife, Evelyn, his mother, Rose Lawson, and his young children, Robert and Diane, were on their way to the Isle of Palms via the Grace Bridge when the collision occurred. The impact sent their 1940 dark green Oldsmobile plunging through the gap in the bridge and into the cold waters of the Cooper River. Crewmen aboard the *Nicaragua Victory* saw the car go off the bridge and rushed to look for survivors. They observed a large bubble of air where the car splashed into the water. Several minutes later all that remained was floating debris and a large oil slick. For days it was believed that a second vehicle had fallen from the damaged structure, but when the wreckage was salvaged one month later, only the Lawsons' sedan was found, and all five bodies were recovered.

The prophesy of Charles K. Allen from 1929, in which he boasted of the bridge's ability to survive a ship collision, proved sadly off the mark, as the Grace Bridge was

6.2 *Photograph taken shortly after the* Nicaragua Victory *collided with the Cooper River Bridge, on February 24, 1946. The ship carved a 240-foot gap in the roadbed, rendering the bridge impassable. Five members of the Lawson family died when the collision sent their car plunging off the bridge.* Courtesy of Sandy Knisley, from the Robert Legaré Coleman Collection

6.3 *The* Nicaragua Victory *collision.* Courtesy of Sandy Knisley, from the Robert Legaré Coleman Collection

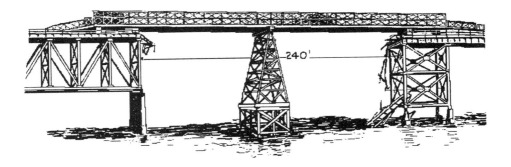

6.4 *A Bailey span used as a temporary patch over the damaged sections of the Cooper River Bridge allowed motorists to utilize the bridge during repair work.* Illustration from the Charleston *News and Courier;* Collection of Pamela Gabriel

not spared major damage.[4] The *Nicaragua Victory* knocked out a massive portion of the roadbed, rendering the bridge impassable. Once again, Charleston was left isolated from areas north and east. Until repairs were made to the span, traffic was rerouted inland, an eighty-mile trip from Mount Pleasant to Charleston. Ferry service was again established. In April 1946, the Army Corps of Engineers installed a temporary, 310-foot long structure known as a Bailey span to patch the bridge's missing section and to restore traffic flow across the bridge. The Bailey span was only wide enough to allow traffic in one direction. Bridge tenders were positioned on either end of the structure to control the flow of traffic. As the Bailey span also had a weight limit of only twelve tons, large trucks and buses were required to cross the harbor via the long, inland route. Permanent repairs were completed in time for the celebration marking the lifting of the tolls in June 1946. The federal government, which had the *Nicaragua Victory* under charter at the time of the accident, paid the $300,000 cost of repairing the bridge.

The accident did not stifle development plans or thwart Charleston's march towards prosperity, but it did magnify Charleston's dependency on a single Cooper River crossing. The point was not lost on area leaders, who began murmuring for a second Cooper River bridge. Ironically, though the Cooper River Bridge had been built to funnel pleasure seekers to the Isle of Palms, it was East Cooper residents for whom the bridge had become a vital part of their daily routine. The shift in public

*6.5 The Bailey span was installed two months after the* Nicaragua Victory *collided with the bridge; until then, motorists were transported across Charleston Harbor in ferries.* Courtesy of the Charleston County Library

perceptions of the bridge and its purpose was significant, reflecting the rapidly changing demographics of Charleston and its suburbs.

## Growth of East Cooper

> *"Mount Pleasant is neither a mount nor is it pleasant . . . "*
>
> —John P. Grace

Until the middle of the twentieth century, the area east of the Cooper River was sparsely populated, much as it had been for centuries. The earliest inhabitants of the area were members of the Wando and Sewee Indian tribes. Some of the first white settlers were New Englanders who arrived in Carolina in 1696, just a few years after the establishment of Charles Town on the peninsula across the Cooper River. On the banks of the Wando River these early colonists established the small settlement of Wappetaw, of which nothing now remains.[5] Between the Wando River and the Atlantic Ocean, on land grants dating back to the Lords Proprietors, large working plantations were laid out. Settlements also sprang up around the several shipyards along the Wando River; one of the earliest of these was located at Hobcaw Point. Much of the land that became Mount Pleasant was originally part of the 2,340 acres granted to Captain Florence O'Sullivan, the same man who later manned the outlook on the island that would bear his name.

Maps of the time indicate the names of landowners, and the hamlets that grew up on their lands and bore their names, such as Barksdale Point and Haddrell's Point. John Barksdale purchased his three hundred acres from Captain Sullivan's daughter in 1694. James Hibben purchased Jacob Motte's plantation, Mount Pleasant, from Motte's estate in 1803. Hibben divided the land into thirty-five building lots and named his planned village after the plantation Mount Pleasant. Years earlier, Hibben's father, Andrew, had purchased a parcel adjacent to Motte's plantation and there established the first direct ferry service between the area and Charleston in 1770. The hamlet that developed at the mouth of Shem Creek around the ferry wharf became known as the Ferry Tract.

William Hort's nearby plantation, Hort's Grove, can still be identified by the oak trees in the park across from modern day Alhambra Hall in Mount Pleasant. Sandwiched between the Motte and Hort estates were one hundred acres belonging to Englishman Jonathan Scott, who laid out the first planned settlement in 1766 as the town of Greenwich. Scott placed fifty town lots on fifty acres along the harbor and left the remaining back fifty acres as commons for the gathering of firewood by town residents. The early settlement of Lucasville along Shem Creek thrived around the rice and saw mills of another English immigrant, Jonathan Lucas. Lucas had purchased Scott's Greenwich Mill at auction in 1793 following Scott's death. Charles Jugnot and Oliver Hilliard created Hilliardsville in 1847 on part of Hort's former lands. They also built the first Alhambra Hall around a picnic area and brought visitors from Charleston to this recreation area on their new ferry service. They filled in some of the marsh areas and promoted the sale of building lots. Gradually all of these small settlements merged and were subsumed under the name Mount Pleasant: Greenwich in 1837, Hilliardsville in 1858, and Hibben's Ferry Tract and Lucasville in 1872. The lands of these early settlements now constitute the Old Village of Mount Pleasant.

Following the Civil War, many of the plantations in Christ Church Parish were broken up, and the land was sold to the former slaves who once worked it. The communities of Remley's Point, Scanlonville, Phillips, Pineview, Snowden, Bee Hive, and Liberty Hill were established as settlements by and for freedmen. Many of the communities thrive today and have maintained their distinctive identities despite the encroachment of modern development.

Mount Pleasant, the largest town east of the Cooper River, grew slowly. Twenty years after the Civil War, Mount Pleasant had a population of fewer than 800 residents. In 1883 Charleston County was divided, and Mount Pleasant became the county seat for Berkeley County. A county courthouse was built on the corner of Pitt and Bank Streets. Six years later county lines were redrawn, and Mount Pleasant

*6.6 Postcard showing an aerial view of the Cooper River Bridge (circa late 1950s) looking across at Mount Pleasant, at the time undeveloped land.*
Collection of Pamela Gabriel

was returned to Charleston County. Though the town lay directly across the harbor from Charleston, residents of the little village of Mount Pleasant were a largely self-sufficient community. The residents, both black and white, earned their living through farming, shrimping, and boat building. High school students attended tuition-free Charleston schools, made possible by a county-city arrangement. Ferries brought Mount Pleasant residents to the city for shopping, medical needs, and entertainment. Various cross-harbor ferry services remained in operation until 1939.

When the Cooper River Bridge opened in 1929, the town of Mount Pleasant had a population of 1,400, and another 1,300 persons resided in the outlying area of Christ Church Parish.[6] Throughout the Depression and the era of the Cooper River Bridge under tolls, the population of Mount Pleasant held steady. The 1940 census indicated a population of 1,700.[7] Even by 1950, four years after the Cooper River Bridge was made toll-free, the population of Mount Pleasant had inched upward to only 1,857. New housing was being developed in Mount Pleasant, especially in unincorporated areas, and new residents were beginning to trickle into town, but a full population boom in East Cooper had not yet been realized.

The Cooper River Bridge affected East Cooper in ways other than spurring new development. Historian Dale Rosengarten credits the bridge with saving the African American tradition of sweetgrass basket weaving. "By 1930, the basket makers had developed a strategy for selling directly to tourists. . . . The paving of [U.S.] Highway 17, formerly Highway 40, and the construction of the Cooper River Bridge turned the coast route that passes through Mount Pleasant into an efficient north-south artery. Basket weavers began displaying their wares on the road and the basket stand was born."[8] Even today, despite setbacks as a result of explosive development in the area, basket weavers still sell their highly valued sweetgrass baskets from stands along U.S. Highway 17.

The transformation of Mount Pleasant from a farming village to a middle-class suburb began in the 1950s. The 1960 census showed that Mount Pleasant's population had surged to 5,116, an increase of nearly 200 percent since 1940. Mount Pleasant became an increasingly younger town, as returning veterans and middle class Charlestonians moved to East Cooper and began families. In 1950, children under the age of fourteen comprised about 32 percent of Mount Pleasant's population, but by 1960 the percentage had increased to 40 percent.[9] The prosperity of the 1950s hit Mount Pleasant with hurricane force as new townspeople began purchasing homes in newly developed subdivisions such as Bayview Acres and the areas surrounding the old village's center. Most of the new housing was located within a few miles of the bridge, and residents could make an easy, toll-free commute to jobs in Charleston and the naval base. New roads and bridges, such as the U.S. Highway 17 bypass, begun in the late 1950s, and the Mount Pleasant Memorial Bridge over Shem Creek, built in 1947 and widened in 1958, helped open new areas of East Cooper to development. Mount Pleasant had become a true bedroom community; however, the growth in East Cooper was not limited to Mount Pleasant. Between 1950 and 1960, the non-military, year-round population of the town of Sullivan's Island increased by 450 persons, or nearly 50 percent, despite a significant decrease in population following the closing of Fort Moultrie in 1947.[10] During the same decade, the Isle of Palms underwent perhaps the most dramatic changes.

The economic depression that immobilized the country during the 1930s put an end to the development plans for the owners of the Isle of Palms. The Hardaway Contracting Company had assumed control of the property in 1934, but it did little to improve or promote the resort. On January 30, 1938, a fire destroyed the island's pavilion, restaurant, and bathhouses. Writing about his visit to the sea islands in 1939, William Oliver Stevens described the summer cottages on the Isle of Palms as "shacks." "They are certainly not inviting as hot-weather retreats,," he continued, "for they look as if they must bake like ovens in the summer sun."[11] Curtis Carter of St. Simons, Georgia, recalls that his grandfather, Dr. King Milligan of Augusta, Georgia, bought "Vacationin," an island cottage, in 1918. "He bought his oceanfront home on the 'island' for the grand sum of $1,800 from Santo Sottile of Charleston," says Carter, who spent many August summers at the cottage during the 1940s and recollects that, contrary to Stevens's observation, these cottages were very cool due to their construction on pilings and the space between the walls that circulated air throughout the house.

Carter remembers crossing the Cooper River Bridge on the way to the Isle of Palms:

> It was such a thrill to see the huge iron Cooper River Bridge looming on the horizon. . . . For a six-year-old, it was just like a roller-coaster ride, all the way to the tickle in my stomach as we went over the top and down again. . . . I just didn't see how we could miss hitting the side. . . . I was wide-eyed, gawking and trying to take it all in, gazing out at Charleston Harbor . . . while Grandaddy gripped the steering wheel in a study of intense concentration and said, 'No talking!' My gosh, this bridge was just about the highest thing in the world. I almost hated for the ride to be over.[12]

J. C. Long, a prominent Charleston attorney, acquired the island in December 1944. Long, operating through his newly founded Beach Company, began to transform the Isle of Palms from a neglected vacation destination into a planned, year-round, residential community. He improved the roads, leveled the sand dunes, cleared the brush and built low-cost housing.[13] The Beach Company also provided necessary services to the community, such as water, trash collection, and street lighting.

Former Isle of Palms mayor Carmen Bunch arrived on the island in 1945 as the bride of a wartime soldier. At the time there were only ten year-round families living

*6.7 Summer cottage "Vacationin" typical of those found on Isle of Palms in the early 1900s. Curtis Carter, whose family purchased the cottage on the Isle of Palms in 1918, remembers crossing the Cooper River Bridge on the way to the beach as a young boy.* Courtesy of Curtis Carter

on the island, including her husband's family, who operated a service station near the inlet. "It was quite an adjustment for a New York girl," she recalls. When she questioned J. C. Long about some of the small, flat-roofed concrete-block houses he was building, he responded that he was building houses for retired people and for those who couldn't afford larger homes but wanted to be near the beach. Some of these homes still exist on the island.[14]

Long's efforts to develop the Isle of Palms were facilitated by improvements to bridges and linking the island to the mainland. In 1945, the old Cove Inlet Bridge, also known as the Pitt Street Bridge and located between Mount Pleasant and the southern tip of Sullivan's Island, was replaced by the new Ben Sawyer Bridge. The new bridge connected to the midpoint of Sullivan's and allowed quicker access to Isle of Palms. The old Breach Inlet trolley bridge linking Sullivan's Island and the Isle of Palms was improved to accommodate automobiles in the 1920s. In 1960 a new concrete-and-steel bridge, Thomson Memorial Bridge (named for Revolutionary War hero William Thomson), replaced the old Breach Inlet bridge. In 2001, work began on a replacement to the 1960 bridge. The new, larger bridge is to be named for the submarine *H. L. Hunley*, commemorating that historic vessel's ill-fated departure from the inlet on February 17, 1864, to sink the Union ship *Housatonic*, blockading Charleston Harbor.

Slowly the Isle of Palms community began to grow, and on May 21, 1953, the island was incorporated as the City of Isle of Palms. By 1960, 1,186 persons were living in Long's planned community.[15] By developing the Isle of Palms into a thriving residential community, Long achieved in spirit what the Cooper River Bridge and its owners had first set out to do, to renew and revitalize the island. Nearly thirty years after the Cooper River Bridge opened the Isle of Palms was a viable beach community.

## Mishaps and Memories

As East Cooper grew, so too did the traffic on the Grace Bridge. By 1959, nearly 12,000 vehicles crossed the bridge each day—more than double the traffic on the bridge five years earlier. Nearly 7,000 new residents moved into East Cooper during the 1950s, and the narrow, two-lane Cooper River Bridge (officially known as the John P. Grace Memorial Bridge, but still referred to as the "Cuppa Riva Bridge" by locals) was one of Charleston's most congested roadways. In addition to the increase in traffic, automobiles had grown larger and wider than those vehicles that first crossed the bridge in 1929; but still the bridge continued to serve two-way traffic. Accidents became increasingly common. As early as the late 1940s, the *News and Courier* began to report on the congestion and traffic on the bridge, and the first rumblings to build a second Cooper River crossing surfaced. The newspaper began a campaign to ease traffic on the Grace Bridge. In a series of articles and editorials it urged motorists to obey traffic laws and to maintain their vehicles to avoid stalls and breakdowns. As the traffic worsened, the paper changed its calls for safety to an outright demand for another bridge. In 1950, the paper's evening edition editorialized, "With the traffic over the bridge continuing to increase the need for another span over the Cooper becomes more urgent."[16] Between 1950 and 1957, nearly 70 people were hospitalized for injuries suffered in accidents on the Cooper River Bridge. Traffic fatalities were few, largely due to the fact that traffic on the bridge often moved at a crawl as hundreds of cars clogged the narrow roadway.

Lifelong Mount Pleasant resident James Craven was a student at Moultrie High School (now Moultrie Middle School) and vividly remembers the excitement of driving across the Copper River Bridge:

> I was a high school student and drove a school bus to school sporting events. Crossing the bridge was a challenge, especially when we met up with a logging truck or a gasoline tanker truck. We had to fold the bus mirrors in and sideswipe the bridge railings just to avoid being hit. The truck did the same thing. Even then we lost our mirrors half of the time. I remember that when the school mechanic came to Moultrie to service the buses, he always brought along a dozen extra mirrors to replace the ones we lost on the bridge.[17]

The East Cooper building boom had just begun, and traffic would only worsen. In 1960 the state Highway Department enlarged a 180-foot section of the Cooper River Bridge by 28 feet, creating a bypass intended to be used as a place for motorists to stop for emergencies without clogging traffic. Locals joked that it was also a good place for highway patrol officers to wait for speeding vehicles.

With steep grades, turns, and heavy two-way traffic, the Cooper River Bridge alarmed many motorists. From the day it opened the bridge was branded a thrill ride, and there are countless stories of motorists frightened by the crossing. In 1952, an incensed New Yorker, called a "critical Yankee" by the *News and Courier,* wrote the *New York Mirror* to ask, "Why doesn't Charleston, S.C., do something about that terrible apology for a bridge over which tourists from the North must drive? It kills the Ocean Route because few tourists ever want to drive over this bridge twice; it is a high roller coaster affair, narrow and frightening."[18] The *News and Courier* suggested, partly in jest, that Charleston's Chamber of Commerce set up "a corps of volunteer chauffeurs" to drive people over the bridge.[19] The plan never materialized, but Charleston police were occasionally called upon to escort frightened travelers over the bridge.

Some visitors simply refused to cross the bridge, opting to drive the eighty-mile inland route from Mount Pleasant to Charleston. On occasion southbound tourists would stop at Moultrie High School, situated on old Highway 17 in Mount Pleasant, and hire students to chauffeur them across the bridge. A student driver could expect to be offered two to three dollars for this service, and once the tourists were safely across on the Charleston side, the grateful student would simply hitchhike back.

Former president of the College of Charleston Alex Sanders remarked in 1996, "My primary recollection of the bridge has to do with my old friend, the late Gedney Howe, a renaissance man and a true gentlemen in every sense. Gedney was not afraid of the devil and certainly never feared no man and no thing but one. He was afraid to cross the Grace Bridge. This subsequently complicated his life, especially when he had to travel from his home in Charleston to Pawley's Island. Look at a map. Its almost impossible to get from Charleston to Pawley's Island without crossing the bridge."[20]

Longtime Charleston resident Claire Robinson remembers, "When my mother and her friend would drive from Boston to Charleston to visit me, her friend would call the police to escort her over the bridge. She would put a blanket over her head and not come out from 'under' until she was assured they were over the bridge."[21]

Humorous bridge tales are plentiful. In 1934, a Charleston police officer offered to jump from the bridge for $150. The paper reported, "one woman—evidently not considered among his admirers—said it would be worth $150 to get rid of him."[22] In 1951 Charleston police arrested a sailor's wife as she tried to run over the bridge in an attempt to wave goodbye to her husband's ship. In 1953 a man was arrested for

*6.8 1958 map showing proposed second Cooper River Bridge and U.S. Highway 17 bypass (in dashed lines) around Mount Pleasant's new developments.* Courtesy of Charleston County Library

walking on the edges of the bridge's narrow railings. In 1955, dance instructors offered a city policeman free lessons if he would chauffeur them across the bridge.

## The New Cooper River Bridge

As traffic and traffic accidents increased throughout the 1950s, sporadic cries for a new Cooper River crossing became an orchestrated chorus. Charleston's political and business leaders lobbied state and federal transportation officials in an effort to have the project made a top priority. In 1957 the state Highway Department admitted that the Cooper River Bridge was "becoming more and more inadequate for present-day and future traffic needs" and began planning for a second crossing.[23] In 1959, state officials formally committed to building a new Cooper River bridge. Plans for a second bridge took years to complete, and, unlike the first Cooper River Bridge, the new crossing was paid for with state and federal highway funds. Gone was the era of private toll bridges.

Work on the new Cooper River bridge began in May 1963, with a groundbreaking ceremony hosted by the East Cooper Promotion Association held on May 2. Charleston mayor J. Palmer Gaillard Jr. gave the opening address, titled "It's a Great Day"; longtime chief commissioner of the South Carolina Highway Department Silas N. Pearman conducted a ceremonial driving of the first pile. (The new bridge would bear his name when it opened in 1966.) Festivities included a flyby salute by aircraft from the Charleston Air Force Base, an aerobatics and precision flying demonstration, and a tour of Fort Moultrie. The day ended with cocktails and dinner at Alhambra Hall in Mount Pleasant.

Compared to the Grace Bridge, the Pearman Bridge took twice as long to construct and cost more than twice as much, yet at a final price of $14.5 million the project was completed $3 million under budget. Three men died during construction of the Pearman bridge: Henry Padgett of Islandton, South Carolina, William Thompson of Shreveport, Louisiana, and Sidney Wilson of Camp Hill, Pennsylvania.

Work on the project was divided among fourteen different contractors, a logistical nightmare designed to save costs but which also earned the bridge special mention in the nation engineering publications. The new structure, immediately nicknamed the "new Cooper River Bridge," was also a cantilevered truss, and its design closely mirrored the unique contour of the older Grace Bridge though the Pearman Bridge was constructed to meet prevailing standards as set forth by the federal government. With sufficient government funding, the Pearman Bridge was configured without a dramatic dip over Drum Island, the cost-saving feature of the Grace Bridge that made that structure infamous. The bridge was designed with 5 percent grades and three twelve-foot-wide lanes, with one lane designed to be reversible to allow traffic flow in either direction.

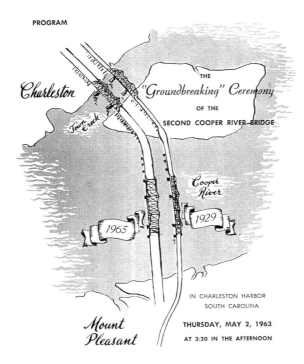

*6.9 Program for the ground-breaking ceremony of the second Cooper River Bridge, May 2, 1963. Planning for the second bridge began in the late 1950s. Charleston attitudes towards the second Cooper River Bridge were markedly different from the city's reception of the first bridge. While the "old" bridge was greeted as an engineering marvel and a great boon to the city, the second bridge was seen primarily as a long-overdue means to handle traffic.* Courtesy of the Charleston County Library

When the Silas N. Pearman Bridge opened in 1966, Charleston's celebration was subdued, in stark contrast to the days-long celebration organized for the opening of the Grace Bridge. The opening of the Pearman Bridge was overshadowed by world happenings. On opening day the bridge shared the front page of the local newspapers with the latest news from the war in Vietnam. The sword of Confederate General P. G. T. Beauregard, which was intended to be used to cut the ceremonial ribbon to open the Pearman Bridge, was kept locked in City Hall for fear the weapon would hurt someone at the ceremony. Instead, ordinary scissors were used for the job. In 1929, the Grace Bridge was heralded as a "bridge of firsts," its opening the dawn of an era for a new, modern Charleston. The Pearman Bridge was called a "long overdue" bridge. Perhaps traffic-weary Charlestonians had grown too sophisticated to appreciate the novelty of a new bridge. The new structure was certainly no superlative, and it lacked the luster that surrounded its predecessor. To wit, among the noteworthy "firsts" for the Pearman Bridge was the fact that it was the first bridge in South Carolina to use pre-stressed concrete in its construction, a milestone hardly worthy of celebration.

Though Charlestonians may not have appreciated the Pearman Bridge's engineering or historical context, they certainly welcomed its immediate impact on traffic. By

*6.10 The Silas N. Pearman Bridge, with the older John P. Grace Memorial Bridge in the background. The Silas N. Pearman Bridge, known as the New Cooper River Bridge, was opened in 1966.*
Photograph by Bill Morgan, Charleston, S.C.

relieving the significant traffic bottleneck between Charleston and Mount Pleasant, the new bridge facilitated unprecedented growth of Mount Pleasant. Between 1960 and 1970, the population in East Cooper increased from 5,116 to 6,879. By 1980 the population of Mount Pleasant grew to 14,209. The town's population had more than doubled since 1970 and nearly tripled since 1960.[24]

The 1970s also brought dramatic changes to the Isle of Palms. In 1975, the Sea Pines Company acquired a 900-acre tract on the northeast end of the island and began plans for an elaborate beach resort, called the Isle of Palms Beach and Racquet Club. In 1984, the resort was renamed Wild Dunes Beach and Racquet Club. With golf courses, homes, and other amusement facilities, the resort is hauntingly similar to the grand plans formulated by the owners of the Cooper River Bridge in the mid-1920s.

## The Bridge Run

Driving over the Cooper River bridges has never been a task for the fainthearted. In 1978 a new challenge was offered: crossing the Cooper River Bridge by foot as part of the first Cooper River Bridge Run. The race was the dream of Dr. Marcus Newberry of the Medical University of South Carolina in Charleston. In an effort to encourage and promote heath and fitness for the community, Dr. Newberry hit upon the idea of a high-profile event, a run across the Cooper River Bridge. Working with officials from the city, local colleges, and the running community, Dr. Newberry organized the first race. The morning of the first run—Sunday, April 2, 1987— was pure Charleston: 82 degrees and muggy. The entry fee was three dollars, which included a T-shirt. Race organizers anticipated 500 runners, but to their astonishment nearly 800 participants appeared the morning of the race. The turnout overwhelmed the committee, and they quickly ran out of official identification numbers worn by runners and began improvising with slips of papers.

*6.11 Runners and walkers cross the Silas N. Pearman Bridge as part of the annual Cooper River Bridge Run. The run, a ten-kilometer footrace first held in 1978, has become one of the largest in the world, attracting nearly 30,000 runners and walkers. During the early years of the race, runners crossed the John P. Grace Memorial Bridge, but by 1995 the race had outgrown the old bridge and the course was moved to its current route over the newer Silas N. Pearman Bridge.* Photograph courtesy of Rick Rhodes, Cooper River Bridge Run

*"As a high school student, I ran the Cooper River Bridge Run, and finished in the top twenty. I remember seeing open road between me and the end of the bridge, with just a few runners in front of me. Now a faster crowd always separates me from the finish line, and I am annually reminded of the passing of my youth."*

—Mark Sanford, U.S. Congressman, March 7, 1996

## 6.12  Cooper River Bridge Run: A Charleston Legend

| Year | Official Entrants | Temperature at Race Start (°F) | *Male* Winner. Home, Finishing Time (minutes:seconds) | *Female* Winner. Home, Finishing Time (minutes:seconds) |
|---|---|---|---|---|
| 1978 | 1,040 | 82 | Benji Durden, Georgia, 30:22 | Lisa Lorrain, Georgia, 39:39 |
| 1979 | 1,350 | 60 | Avery Goode, South Carolina, 32:55 | Marty Long, Summerville, S.C., 40:10 |
| 1980 | 1,500 | 59 | (tie) Kim Burke & Steve Littleton, Pennsylvania, 31:26 | Michelle Moore[2], 41:29 |
| 1981 | 1,650 | 60 | Marc Embler, Columbia, S.C.[1], 30:54 | Kiki Sweigary, Connecticut, 35:10 |
| 1982 | 2,100 | 45 | Mark Donahue, Charleston, S.C., 30:28 | Sallie Driggers, Hanahan, S.C., 37:21 |
| 1983 | 3,115 | 50 | David Branch, South Carolina, 29:28 | Mary Copeland, Charleston, S.C., 38:09 |
| 1984 | 4,459 | 50 | David Branch, South Carolina, 29:25 | Brenda Webb, Tennessee, 34:09 |
| 1985 | 5,440 | 70 | Mike O'Reilly, Ireland, 29:28 | Christina Boxer, England, 34:08 |
| 1986 | 6,684 | 72 | Hans Koeleman, Netherlands, 29:29 | Lesley Welch, Massachusetts, 33:37 |
| 1987 | 6,997 | 39 | Paul Cummings, Utah, 30:19 | Mary Ellen McGowan, Florida, 34:31 |
| 1988 | 6,904 | 65 | Ashley Johnson, Kentucky, 29:56 | Carla Borovicka, Florida, 34:38 |

[1]A student at the University of South Carolina from the Charleston area.
[2]Hometown not known.

| Year | Official Entrants | Temperature at Race Start ( °F ) | *Male* Winner. Home, Finishing Time (minutes: seconds) | *Female* Winner. Home, Finishing Time (minutes: seconds) |
|---|---|---|---|---|
| 1989 | 7,510 | 55 | Ashley Johnson, Kentucky, 29:48 | Grete Waitz, Norway, 33:29 |
| 1990 | 7,820 | 50 | Sam Obwacha, Kenya, 29:20 | Shelly Steely, Oregon, 32:57 |
| 1991 | 6,527 | 64 | Jeff Cannada, North Carolina, 29:38 | Kim Bird, Georgia, 34:49 |
| 1992 | 7,602 | 48 | Dominic Kirui, Kenya, 28:24 | Jill Hunter, England, 32:34 |
| 1993 | 7,544 | 50 | Paul Bitok, Kenya, 28:31 | Sabrina Dornhoefer, Minnesota, 33:53 |
| 1994 | 8,670 | 60 | Simon Karori, Kenya, 28:35 | Elaine Van Blunk, Pennsylvania, 34:01 |
| 1995 | 12,407 | 59 | Joseph Kimani, Kenya, 27:49 | Laura LaMena-Coll, Oregon, 33:58 |
| 1996 | 14,030 | 50 | Joseph Kamau, Kenya, 28:32 | Liz McColgan, Scotland, 31:41 |
| 1997 | 15,216 | 68 | Paul Koech, Kenya, 27:57 | Elana Meyer, South Africa, 31:19 |
| 1998 | 18,007 | 64 | Tom Nyariki, Kenya, 29:58 | Elana Meyer, South Africa, 32:46 |
| 1999 | 15,349 | 45 | Lazarus Nyakeraka, Kenya, 28:40 | Eunice Sagero, Kenya, 33:18 |

*6.12  Cooper River Bridge Run: A Charleston Legend (continued)*

| Year | Official Entrants | Temperature at Race Start (°F) | *Male* Winner. Home, Finishing Time (minutes: seconds) | *Female* Winner. Home, Finishing Time (minutes: seconds) |
|---|---|---|---|---|
| 2000 | 16,893 | 61 | James Koskei, Kenya, 27:40 | Catherine Ndereba, Kenya, 31:42 |
| 2001 | 16,884 | 64 | James Koskei, Kenya, 28:45 | Catherine Ndereba, Kenya, 32:33 |
| 2002 | 16,719 | 51 | John Itati, Kenya, 28:06 | Catherine Ndereba, Kenya, 31:53 |

The first running of the Cooper River Bridge Run began at Patriot's Point in Mount Pleasant, wound over the single reversible lane of the Pearman Bridge, and ended at historic White Point Gardens on the Battery in Charleston. Since state law prohibited use of the bridge for foot traffic, special legislation had to be written that allowed the bridge to be closed for one day to accommodate the run. This first race was not without problems: water was not provided, and given the heat and humidity, numerous runners suffered heat exhaustion. Despite this, the race was held again the following year, and even more runners participated. In 1979, the course was redirected over the older Grace Bridge. The race ran over the old bridge from 1979 until 1995. The bridge's steep dips, curves, and narrow lanes added to the challenge and lore of the race. As thousands of runners crowded onto the bridge, the impact of foot strikes caused the bridge to sway—an uncomfortable (albeit structurally safe) feeling for many. Each year the Cooper River Bridge Run grew in size. By 1995, annual participation in the event had surged to more than 15,000 participants and had grown too large for the old Cooper River Bridge and its narrow roadway. Since 1996, the race has been run over the Pearman Bridge.

The official entry application for the first run carried the prophetic line: "destined to become a legend." The Cooper River Bridge Run has become legend; it is one of the largest ten-kilometer (6.2 miles) footraces in the world, and in recent years it has drawn nearly 30,000 runners and walkers. Today the race hosts runners from nearly every state and many foreign countries and is the largest tourist weekend in Charleston, generating an estimated $9.3 million in economic impact.

## The End of the Bridges

Early in the 1900s, as the automobile changed the nation's method of travel, highways and bridges were erected to accommodate the new mode of transportation. Today these structures are nearing the end of their functional lifespans. In a sense, the fate of the Grace Bridge in modern times is no different from that of countless other bridges across the country. Neglect, overuse, and perhaps inadequate design and construction have forced the replacement of hundreds of old structures, precipitating what some have called the second great transportation building boom.

Since 1929, the Grace Bridge has shouldered the burden of East Cooper's tremendous growth. Decades of heavy traffic and, more importantly, years of neglect and poor maintenance have advanced the deterioration the bridge. Cleaning and painting have been skipped or postponed, and Charleston's salty, humid atmosphere has eaten away at the bridge's steel structure. In the late 1980s, the John P. Grace Memorial Bridge received the dubious distinction of being cited as one of South Carolina's most unsafe bridges, based on surveys performed by the state's department of transportation. In fact, in recent bridge safety measures administered by the U.S. Department of Transportation each year, the John P. Grace Memorial Bridge has attained a safety rating of only 4 out of a possible score of 100.

Throughout the 1990s, media attention focused on the bridge's poor safety ratings and its structural soundness. However, the national bridge safety rating was designed to identify those bridges most in need of replacement, and is, at best, a conservative rating of the bridge's structural integrity. The scoring system is based on three factors. First, structural integrity accounts for 55 percent of the overall score. Second, functional obsolescence, which rates how a bridge conforms to current federal highway standards, accounts for 30 percent of the final score. And finally, the importance to public use accounts for 15 percent of the final score. A bridge that is heavily utilized would receive a low score in this category, indicating that it is of great public use and therefore more worthy of replacement.

The Grace Bridge is a functionally obsolete bridge. Based on current regulations, the lanes are too narrow, its grades are too steep, and it lacks certain safety features, such as roadside shoulders, that are now required by federal highway codes. One of the recent comprehensive inspections of the Grace Bridge, performed in 1987, rated the bridge's structural integrity as "fair to poor."[25] The inspection noted that at certain points on the bridge's truss rust had eaten away nearly 75 percent of the original steel. Rust has also frozen some of the bridge's expansion joints, components that allow the steel to expand and contract with changes in temperature, rendering them inoperative and thereby increasing stress on other steel components. In his book *Bridge Engineering*, Dr. Waddell commented that a bridge, if properly constructed and maintained, would last forever. Over the years the Cooper River Bridge has been plagued by poor maintenance. Regular applications of paint could have arrested much of the deterioration to the bridge's steel. While the bridge was under private ownership, from 1929 to 1941, routine maintenance on the structure was skipped or delayed in order to cut operating costs The debt-burdened bridge owners were reluctant to undertake expensive maintenance projects, and even routine upkeep was costly. For example the cost to paint the bridge in 1935 was $30,000—nearly half the bridge's annual operating budget—and required almost eighteen months and 6,000 gallons of paint to complete.[26] The bridge under state ownership has fared little better. The state has occasionally delayed expensive repainting, averaging only one complete painting of the bridge per decade. Even when the structure has been painted, inspections have found that the work was faulty and of little use. Years of dirt and debris that should have been removed prior to repainting have simply been painted over and have contributed to bridge's deterioration.

Since the 1987 inspection, emergency repairs have been made to correct the most serious of the bridge's problems. State transportation officials have also taken the precautionary step of prohibiting large trucks and buses from crossing the Grace Bridge. These actions offer little comfort to area residents concerned about the bridge's safety. As early as the late 1970s Charlestonians were beginning to publicly voice their concerns about the future of the two Cooper River bridges and the Grace Bridge in particular. In 1979, the *News and Courier* hired independent engineering consultants to inspect and analyze the bridge. The opinion of the consultants stated that bridge was heavily overstressed and overused, and that it should be closed immediately. Inspections of the structure performed by the state and its consultants refuted these claims. Nevertheless, Charlestonians have remained jittery over the bridge's structural integrity. Late in 2000, a bridge inspection revealed a possible crack in a steel beam and the Grace Bridge was immediately closed to all traffic. Further inspection determined that the damage was rust and not a crack, and the bridge was subsequently reopened. Charleston Mayor Joe Riley was the first to ride across the bridge, to demonstrate its safety.

In the 1990s people continued to move into the East Cooper area despite the horrendous effects of Hurricane Hugo in 1989. It was apparent that Mount Pleasant was one of the fastest-growing communities in South Carolina, with a population of over 30,000. By the turn of the millennium the population of the "town" approached the 50,000 mark. Despite its being labeled as unsafe and inadequate, old residents and newcomers alike continued to rely upon the Cooper River Bridge, fighting the daily congestion between Charleston and Mount Pleasant. Prior to the opening of Interstate 526 in 1989 the average daily traffic for both spans was 64,500 vehicles.[27] While accidents continued to be frequent occurrences on both bridges, fatalities were rare. In 2001, dividing posts were placed along the reversible lane of the Pearman Bridge following an increase in the number of accidents caused by cars swerving into the path of oncoming traffic. The speed limit was also reduced to 40 miles per hour in a further attempt to reduce the accident rate.

Suicide attempts have occurred. The bridges—like all tall structures—possess an attractive lure for desperate souls seeking to end their lives. While exact figures have not been recorded, a 1979 *News and Courier* article stated that there had been fourteen suicide attempts on the bridges; in four cases, the jumper survived.[28] Since the Pearman Bridge was constructed, it has attracted the greater number of jumpers, perhaps due to its pedestrian walkways. This gruesome aspect of the bridges' history has found its place in fiction as well: in the opening paragraph of Pat Conroy's 1995 best-seller *Beach Music*, the wife of the novel's main character leaps to her death from the new Cooper River Bridge.[29]

The daily frustrations of dealing with traffic congestion on the Cooper River bridges caused many local residents to lose both their fondness for and appreciation of the two crossings. However, the bridges continued to fascinate writers and artists. In paintings and drawings, the iron spans loom majestically above the rooftops and spires of the old city. Photographers delight in the silhouette of the bridges at sunset. Southern writers make metaphorical use of the bridges to illustrate a character's escape from oppression or as a route home for the prodigal son. In Pat Conroy's *The Prince of Tides* protagonist Tom Wingo rejoices, "Behind me the city of Charleston simmers . . . and before me my wife and children are waiting for me to arrive home. . . . As I reach the top of the bridge . . . I drive toward my Southern home."[30] In her 1999 best-seller *Sullivan's Island*, Dorothea Benton Frank's main character, Susan Hayes, escapes from the emotional turmoil of her Charleston marital home and travels via the bridge to seek sanctuary in the island of her childhood: "As my car swung around East Bay Street to take the bridge east of the Cooper, I found my heartbeat slowing down. I was on my way back to the beach—the cure-all for whatever ails me."[31]

Screenwriters and advertisers have also featured the bridges' unique forms. In the 1995 action thriller *Die Hard with a Vengeance (3)*, the heros portrayed by Bruce Willis and Samuel L. Jackson watch as the villain leaps over the side of the old Cooper River Bridge onto a passing freighter. Undaunted, the good guys fling a cable attached to their truck onto the slow-moving ship. They then slide down the cable onto the deck of the freighter, which dramatically drags the truck over the side of the bridge and sends it splashing into the waters below.

By the early 1980s Charleston leaders were demanding the replacement of both Cooper River bridges. In the mid-1990s state transportation leaders formulated rough plans for a new, ten-lane bridge to replace the existing Cooper River bridges, but balked at the estimated $400 million price tag. State officials demanded that Charleston help pay for the new bridge through a combination of new taxes or even tolls. Charleston leaders reeled at the suggestion. Lowcountry politicians pointed to the

*6.13 Dignitaries gather for groundbreaking of the new, multilane Arthur Ravenel Bridge. The bridge will replace the two existing Cooper River bridges. Seated at the table (from left to right) are U.S. Senator Ernest "Fritz" Hollings, South Carolina Governor Jim Hodges, Mount Pleasant Mayor Harry Hallman, Charleston Mayor Joseph P. Riley Jr.* Photograph by Mark Richardson; courtesy of the *Charleston Regional Business Journal*

*"We rode over it the day it opened. I was seven years old and convinced we would be going over the top of the structure—I sat on the floor of the car! When I was running for lieutenant governor, some of the powers-that-be jumped all over me for fighting the toll on the Cooper River Bridge."*

—Ernest F. "Fritz" Hollings, U.S. senator, July 27, 1998

extensive highway improvement projects in upstate South Carolina and cried foul. In a fight strikingly similar to the bridge purchase debates of the 1940s, Charleston politicians were pitted against upstate interests in a fight to determine who would pay for the bridge. It was yet another chapter in the centuries-old rivalry between upstate and lowcountry interests.

The opening of the Mark Clark Expressway in 1992, which connected Mount Pleasant to North Charleston via the new Don Holt Bridge across the Cooper River, offered considerable traffic relief to East Cooper residents. However, the old Cooper

**Charleston's Grace Memorial Bridge Available**

The South Carolina Department of Transportation (SCDOT) is replacing the bridges over the Cooper River in Charleston harbor beginning in the Spring of 2002. The two-mile historic Grace Memorial Bridge, eligible for the National Register of Historic Places, is being offered for removal and relocation. Designed by Waddell and Hardesty of New York and built by the McClintic-Marshall Company of Pittsburgh, the two-lane bridge opened to traffic in 1929. Proposer must relocate and rehabilitate the bridge, in its entirety or segments, according to the terms of the MOA. Limited funds will be available to defray the costs of removal but will not cover the entire cost of relocation. Method of removal will be controlled by agreements with the U. S. Army Corps of Engineers, U. S. Coast Guard, and others.

*6.14 Want to buy a bridge? Because of the age and significance of the John P. Grace Memorial Bridge, it is eligible to be listed on the National Register of Historic Places. The South Carolina Department of Transportation listed the bridge for sale through advertisements such as this one.*
Used with permission of the South Carolina Department of Transportation

River bridges remained the only direct link between East Cooper and the city of Charleston. The original plan for a ten-lane bridge was determined too costly and new design options were debated. Officials briefly considered replacing only the Grace Bridge, instead of replacing both Cooper River bridges. By 2001, after nearly ten years of debate, state and local officials reached an agreement on a single bridge design consisting of eight lanes to replace both existing bridges. The new structure is to be financed through a combination of state, federal, and local monies at a total cost of over $600 million, more than one hundred times the cost of the first Cooper River Bridge. On July 2, 2001, the groundbreaking for the new Cooper River bridge, named the Arthur Ravenel Jr. Bridge, gathered together federal, state, and local officials and ended the years of controversy. The new bridge will be a cable-stayed design, similar in concept to a suspension bridge but less expensive to construct. Strong bundles of steel cables will be attached to massive, tall piers and will extend down to support the bridge's road deck. The Arthur Ravenel Jr. Bridge will be huge—the largest cable-stayed bridge in North America and the largest construction project in South Carolina history.[32]

Throughout the population explosion of the Charleston area, the one constant has been the Grace Bridge. Despite its age and its current limitations, the bridge has served the communities on both sides of the Cooper River and has transported more than four generations of Charlestonians to and from their beloved city. The bridge has stood firmly, but attitudes toward Charleston's most visible landmark have not. Each generation since 1929 has regarded the Grace Bridge in a different light. In the 1920s, citizens were awed by the bridge's majesty and inspired by the hope for prosperity it represented. In the 1950s a new generation saw the bridge as a route to new suburbs and an escape from congested city life. The next generation cursed the bridge as unsafe and inadequate and demanded a new structure to ease their commutes. Today, a new generation will have the historic opportunity to witness both the construction of the newest Cooper River bridge and the removal of the old bridges. When the older Cooper River bridges are destroyed, Charleston will lose a celebrated landmark; their removal will close another chapter in Charleston's long and colorful history.

# Epilogue

A new silhouette will soon span the Charleston harbor. A sleek and graceful cable-stayed bridge will traverse the Cooper River between the historic city of Charleston and the burgeoning town of Mount Pleasant. The spiny steel arches of the Grace and Pearman bridges, which appear to support each other as they loop across the Charleston skyline, will be replaced by the higher, straighter, and modern Arthur Ravenel Jr. Bridge. The new structure will be massive, dwarfing the old bridges. Its shiny, polished newness will contrast dramatically with the rust and peeling paint of the aging Pearman and Grace structures.

The new bridge will bring much-needed relief to the traffic and congestion of Charleston and its suburbs. No new visions and no grand aspirations are promised with this structure. It will be enormous, it will be impressive, but it will just be a bridge. It will not be the symbol that the Great Cooper River Bridge was for the thousands that looked upon its skeletal form and dreamed great dreams in 1929.

When the old Cooper River Bridge was built in 1929, it was not only a giant leap across the Charleston Harbor, it represented a giant leap of faith for a city bound to the past by pride and tradition. The builders of the old bridge envisioned a place for an antiquated city in a modern rapidly changing century and firmly believed that their bridge was the catalyst for this change. The old bridge fulfilled this promise made to the Charleston community in 1929. Charleston did thrive and prosper, and the East Cooper area did grow. Perhaps, the timetable of these visionaries was premature but the hopes and dreams they imagined for Charleston have been realized. The Great Cooper River Bridge spanned more than a harbor—it stretched across time. It brought a decaying eighteenth-century city into the twentieth century.

For more than seventy years the old Cooper River Bridge has been a towering presence in Charleston. It has served not only as a symbol of the city, with its image imprinted on calendars, T-shirts, and postcards, but it has reflected the shifting times and attitudes of both a city and a nation. It changed from the physical manifestation of the progressive ideas and optimism in the 1920s to one of debt and despair in the 1930s. It witnessed the departure and return of thousands of young men and women during years of fear and war. Later it became an avenue for fulfilling the American

Dream as young persons departed the crowded cities for homes of their own in the suburbs. The old Grace Bridge has become a victim of its own success; yet, it has been as much a symbol of the Holy City as the spires of its churches.

It is time to say goodbye to the Great Cooper River Bridge and applaud its service to the Charleston community. You did well, old bridge. We will miss you.

# APPENDIX

## John Grace, the Man

John Grace was a man much like the bridge that bears his name—he was both reviled and adored. When the bridge named in his memory is no longer a part the city of Charleston, his name will be all but be forgotten.

—————•—•—•—————

*A sliver of the first-quarter moon lingered in the early morning sky. A young bare-foot boy sloshed through ankle-deep puddles. It had rained during the night, and the lowlands of the Charleston peninsula struggled to return to their natural state of marshlands and streams. The rain did little to cool the hot humid air, now heavy with the combined stench of flooded privies, damp hay, manure, and wet barnyard animals. The boy carefully positioned the notched pole across his shoulders, balancing the tin buckets of warm milk, and ventured out into the muddy steamy streets littered with the debris of the city. It was his job to deliver the fresh milk to his neighbors in "the borough." The boy did not complain; it was the least he could do to help his widowed mother as she attempted to support herself and her children.*

Years later newly elected Mayor John Patrick Grace would vividly recall the memory of these early morning milk deliveries. One of the first orders of business facing the new mayor was an ordinance to prohibit the keeping of cows within the city limits and the selling of milk from backyard dairies. Persuaded by local doctors and health officials that these steps were necessary to rid Charleston of periodic typhoid epidemics, especially among the city's poor, Mayor Grace signed the potentially politically embarrassing law, ever mindful of the fact that such an ordinance could adversely affect those individuals whose support helped put him in office. In the *Charleston City Yearbook* of 1911, Grace anguished over this decision in his annual mayor's report. He explained that the "choice [was] between saving babies and sacrificing widows." In an effort to distance himself from the poverty of his youth, he further declared that he was assured that there was no one in the city whose livelihood was dependent on this homegrown industry.

The memories of the harsh struggles of his younger days carried into his adult life in his propensity to fight for those causes near and dear to him, as well as those injustices caused by the inequities of the classes. In 1907 he attracted notice in the case of

*Ex parte Drayton*, a landmark ruling that effectively ended the practice of peonage in South Carolina. This heinous system held many African Americans sharecroppers virtually in bondage to the merchant-planters whose land they farmed. Under the practice, the planter advanced the sharecroppers certain goods throughout the year, often at inflated prices. When the crops were harvested, the sharecropper owed the planter not only rent for the land, but the cost of the goods he received. At the end of the year, these sharecroppers were further indebted to the planter, perpetuating the bondage state. The handling of this case brought Grace the acclamation of presiding District Court Judge William H. Brawley:

> The petitioners in this case are the poorest and humblest class of citizen. . . . Their case has been brought here by a young member of this bar, himself belonging to a race that in the past has suffered through centuries of injustice and oppression, whose heart has been touched by the cry of the lowly, and who apparently at his own cost, from sheer love of liberty and a hatred of wrong makes this appeal for the liberty to which they are entitled under every sanction of the constitution and laws of their country.

Encouraged by these words of praise, the young Grace devoted his entire life to the causes he believed in.[1] Despite his long and successful career as a partner in a prestigious law firm, Grace was never wealthy. In his most successful year, he earned $42,000, a substantial sum in the 1920s, but he diverted much of his resources to his other interests: his political campaigns, his newspaper, the cause of Irish independence, and various charitable works.

Grace's two terms in City Hall were characterized by numerous successes. During his first term (1911–1915), in addition to improving city health regulations, he negotiated with the Seaboard Air Line Railroad for service into Charleston. For years, rail rates into Charleston were disproportionately high, causing planters and farmers throughout the state to choose other outlets besides the Charleston harbor for shipping their goods. By bringing a new line into the city, Mayor Grace was able to secure more competitive rates.

Grace's second term in office (1919–1923) reflected his entrenched beliefs in the Progressive ideals he had acquired during his travels to the North and Midwest in his youth. He strongly promoted municipal ownership of utilities. He championed for municipal purchase of the decaying city docks, which were under the control of the railroads through the East Shore Terminal Company. It was this particular issue that secured his reputation, overshadowing even the unpopular anti-British/pro-German campaign in Grace's newspaper, *The Charleston American*. The subsequent acquisition of the docks created the Ports Utility Commission, the forerunner to the modern day State Ports Authority. Grace was also successful in many of his efforts to pave city roads, and Charleston officials boasted they had the best-paved roads in the South.

When Prohibition became the law of the land in 1919, during Grace's second tern, most Charlestonians simply chose to ignore the eighteenth amendment, much in the same manner that they had disregarded a South Carolina referendum passed in 1893 favoring total prohibition of liquor. It was said that Grace, despite his many connections with the proprietors of the illegal "blind tigers," was a non-drinker. He personally objected to enforcing these laws, since he believed these vices were "sins," not crimes.

His association with individuals of questionable character brought him much criticism. In 1913, his friend, Frank "Rumpty Rattles" Hogan, testified before a congressional investigation into the buying of votes in South Carolina. Grace presented Hogan to the committee as "absolutely trustworthy." Hogan claimed to be a Charleston policeman (actually, he had been appointed to the position eleven days earlier by Grace) and attested to the practice of buying of votes during a Congressional election in which he took part. During the election in question, he campaigned against Grace's candidate, therefore, the unscrupulous actions he observed did not reflect upon Grace. The committee, however, was unimpressed by Hogan and a member later described him as "absolutely unworthy of belief" and offered "no credence whatever . . . in his affidavits or verbal testimony."[2] Always one to seize the limelight, Grace explained to the committee that he had nothing to gain by Hogan's testimony and proceeded to compromise Hogan's reputation by describing Hogan's illegal liquor business. He further added that there were 250 similar establishments in a city with a population of 60,000. The purpose of this revelation by Grace was unclear, but Hogan subsequently remained in the Grace "camp." In 1927 the married and fifty-one-year-old Hogan was shot and killed as he left an East Bay restaurant with his eighteen-year-old girlfriend. One of the alleged killers was Hogan's son-in-law.

In later life, Grace defended a man who confessed to killing his mistress. Grace was successful in saving the man's life by winning a manslaughter conviction for his client, whom he portrayed as an "avenging angel of God," who "shot [the woman] for her sins."[3] Grace could also not resist grandstanding, pointing out to the jury that his client was being tried because he was a poor man, while the crimes of the Bourbons, who stole "thousands of dollars of public money," went unpunished. Reportedly the defendant whispered to Grace that since he was speaking of Bourbon, he would like a "little nip of the stuff."[4] A side note to this story: the convicted man was sentenced to twenty years at the state penitentiary, but was paroled after less then three years and later received a full pardon.

Following his last bid for public office in 1931, Grace was sued for the mishandling of certain property in the estate of Charles O'Dell. A hearing was held on this matter in federal court in 1933 and Grace was absolved of any misconduct. In 1932 the Logan and Grace Law Firm formally dissolved, and while no public statement was issued, it was rumored that a there was a falling out over the 1931 mayoral race in which Burnet R. Maybank was elected. Grace had withdrawn his support for Maybank, who was offered as a nonpartisan candidate and who had the support of the business community. It was also in 1931 that Grace's appointment to the highway commission expired, but he continued to serve at the pleasure of the governor until he was replaced in 1933.

Despite his flamboyant political life, Grace lived a quiet private life. He was devoted to his mother and continued to make his home with her even after his marriage at the age of thirty-eight. His wife, Ella Barkley Sullivan, was a Charleston woman of Irish ancestry. They were married in New York City on November 27, 1912, during Grace's first term as mayor. The couple never had children. Later they resided in a modest house located at 174 Broad Street, where Mrs. Grace invited neighborhood children into their home for cookies and milk. Senator Benjamin Tillman wrote Grace in 1913, "I really want to like you on account of your wife whom I love and who is a noble woman."[5] In the same memo Tillman also advised Grace to emulate his quieter partner, Logan, so that he "would not get into hot water so often."[6]

Grace's remaining years were relatively quiet. Both Grace and his wife were devout Catholics, worshiping at the Cathedral of St. John on Broad Street. Grace belonged to the Knights of Columbus and the Hibernian Society of Charleston (of which he served as president from 1935–1937). In June 1940, at the age of sixty-five, Grace suffered a heart attack. He appeared to be recovering and was released from the hospital and returned home, where he died a week later, on June 25th. State, county, and local politicians attended his funeral service at the cathedral, where he was eulogized by the Right Reverend Joseph O'Brien with the words, "He might have been rich; but riches stain. He might have been powerful; but power corrupts; he cast his lot among the socially unsecure, the weak and lowly."[7] Among those serving as pallbearers were former state senator J. C. Long and Mayor Daniel Sinkler. He was buried in St. Lawrence cemetery, where his tombstone bears the words of Judge Brawley.

# NOTES

## Introduction

1. Elizabeth Mock, *The Architecture of Bridges* (New York: Museum of Modern Art, 1949), 79.

2. John P. Grace, "A Review of the Cooper River Bridge," in "Cooper River Bridge Celebration, Charleston, S.C., August 8, 1929." Available at the *Post and Courier* Library, Charleston, S.C.

## Chapter 1

1. Grace, "A Review of the Cooper River Bridge."

2. The Coastal Highway was also known both as the Old Kings Highway and the Washington Highway. It the late 1920s it became part of Route 40 in South Carolina, and later still it became U.S. Highway 17.

3. John Ficken, an attorney, served in the South Carolina General Assembly from 1876 until 1891, when he was elected mayor of Charleston.

4. "Concrete Structure Dedicated Ten Years Ago Tomorrow Is Vital Road Link and Gateway to Suburbs," *News and Courier*, May 4, 1936.

5. The phrase "An all paved road from Quebec to the Florida Keys" became a national slogan for the U.S. Chamber of Commerce.

6. "Mt. Pleasant Is Optimistic—Mayor McCants Declares Mighty Bridge Dream Come True," *Charleston Evening News*, August 8, 1929. At the opening of the Cooper River Bridge in 1929, Mount Pleasant mayor T. G. McCants claimed that the creation of the McCants Ferry Company (opened by his ancestors) was "a great point in the growth of the town." Census records indicate that Mount Pleasant's population grew very little through the late 1800s and early 1900s.

7. Prior to 1968 the streets were know by various names: Osceola was known as both Cove Street and Main Street. On some earlier maps the numbered streets have names (Frost Street for Station 23, for example). In 1968 the streets were officially renamed and the trolley station numbers reinstated.

8. According to Mary Sparkman's account, *Through a Turnstile into Yesteryear* (Charleston, S.C.: Walker, Evans and Cogswell, 1966), "Parm" derived from a speech pattern peculiar to Charleston wherein certain words like *palm* and *psalm* were pronounced to rhyme with *harm* (49).

9. In his book *A Short History of Charleston* (Columbia: University of South Carolina Press, 1997), historian Robert Rosen states that the unsavory O'Sullivan was known as an "ill-natured buggerer of children" (14).

10. In 1999, Fort Moultrie and the Avery Research Center placed a memorial marker on the site where "Africans were brought to this country under extreme conditions of human bondage and degradation. Tens of thousands of captives arrived on Sullivan's Island from the West African shores between 1700 and 1775."

11. Charles S. Spencer, "Sullivan's Island Is Historic Site and One of the Nation's Oldest Resorts," *News and Courier*, July 21, 1958.

12. The wealthiest Charlestonians summered in northern resorts; Rhode Island was particularly popular with Charleston's elite.

13. The exception to this rule was the New Brighton Hotel (Atlantic Beach Hotel). John Chisholm refused to build the hotel unless he was granted a deed for the land, and the state agreed to his demand.

14. In May 1953, an act of the state legislature gave property holders clear title to their property.

15. "Council Opposes Bridge Plan," *News and Courier*, January 15, 1913.

16. Detailed accounts of the creation and operation of the Cooper River Ferry Commission can be found in the James Allan Papers (1913–1930), South Carolina Historical Society, Charleston, S.C.

17. The wooden Shem Creek bridge was replaced in 1947 with the Mount Pleasant Memorial Bridge, dedicated to the four Mount Pleasant residents who died during World War II. This bridge was widened from two lanes to four in 1958.

18. The bridge burned in 1981. Burning asphalt and creosote pilings made smoke visible for miles. Firefighters could do little to battle the blaze due to the thick smoke, and the bridge was destroyed. The remains of the bridge have been turned into a popular fishing pier, which in 2001 was extended and renamed The "Pickett Pitt Street Bridge and Recreational Area" in honor of the late Dr. Otis Pickett Jr., who died in 1995. Dr. Pickett served the East Cooper area for almost fifty years, crossing the old bridge to reach his patients on the islands (Noelle Orr, "Pitt St. Bridge to Be Named for Dr. Pickett," *Post and Courier*, July 5, 2001).

## Chapter 2

1. South Carolina's infant mortality rate—in the year 2000, 8.7 per 1,000 live births—is still disproportionately higher than that of the rest of the U.S. The South Carolina rate also demonstrates a wide disparity between white (5.5) and minority (14.2) infant mortality ("S.C. Drops to 45th Place in Infant Mortality Survey," *Post and Courier*, 25 May 2002).

2. Artist Elizabeth O'Neill Verner wrote and illustrated a wistful reminiscence of Charleston in the early 1900s titled *Mellowed by Time* (Charleston, S.C.: Bostick and Thornley, 1959).

3. Don H. Doyle, *New Men, New Cities, New South: Atlanta, Nashville, Charleston, Mobile, 1860–1910* (Chapel Hill: University of North Carolina Press, 1990), 181.

4. Ibid.

5. Doyle W. Boggs Jr., "John Patrick Grace and the Politics of Reform in South Carolina, 1900–1931" (Ph.D. diss., University of South Carolina, 1977), 86.

6. How the city lent its name to the dance craze is uncertain. The tune "The Charleston" was composed by jazz pianist James P. Johnson in 1913. In Charleston, the tune was a popular

piece performed by the Jenkins Band, youngsters from a local black orphanage. The dance was first demonstrated in the Ziegfeld Follies in 1923.

7. *City of Charleston Yearbook*, 1929 (Charleston, S.C.: Walker, Evans, and Cogswell), xxix. Mayor Stoney went on to add, "The tourist business has come to be of volume and importance. We will not have Charleston converted into Miami Beach or Palm Beach . . . but we have no objection to the business growing."

8. City zoning regulations, passed after the Francis Marion Hotel was built, state that buildings on Charleston's historic peninsula can be no taller than the district's tallest church spire. Currently, the tallest church spire is that of St. Matthew's Lutheran Church, on King Street.

9. Quotation in issues of *Carologue*, published by the South Carolina Historical Society.

10. "Who Shall Build the Bridge?," *News and Courier*, July 25, 1926.

11. H. F. Church, *City of Charleston Yearbook*, 1930 (Charleston: Walker, Evans & Cogswell), xlii.

12. Ibid.

13. Ibid. Despite Grace's claims in 1929, he does not appear to have played a central role in the bridge project until this time.

## Chapter 3

1. The Judge King Mansion was destroyed in 1938 to make room for the College of Charleston's gymnasium.

2. Most biographical accounts state that Grace "graduated" from Georgetown Law School, but in truth he did not. Up until the 1950s, an individual had only to pass the bar to practice law in South Carolina.

3. John Duffy, "Charleston Politics in the Progressive Era" (Ph.D. diss, University of South Carolina, 1963), 331.

4. *Charleston City Yearbook*, 1911, xvi.

5. Grace to Sydney Anderson, December 5, 1913, quoted in Boggs's "John Patrick Grace and the Politics of Reform in South Carolina, 1900–1931," 45.

6. Ibid., 37.

7. Henry Alexander White, *The Making of South Carolina* (New York: Silver, Burdett and Company, 1914), 217.

8. Duffy, "Charleston Politics in the Progressive Era," 205.

9. Beresford Street was renamed to Fulton during World War II. The renaming was a deliberate attempt by the city to confuse visiting seamen in search of the red light district. On early twentieth-century maps, Archdale is listed as Charles Street.

10. Grace's attitude was not unique; even many of Charleston's genteel ladies viewed these "shady" streets as a picturesque component of the city's diverse tapestry. Elizabeth O'Neill Verner refers to such a street as "Do-As-You-Choose-Alley" in her *Mellowed By Time*.

11. Rosen, *Short History of Charleston*, 143.

12. Grace, "A Review of the Cooper River Bridge."

13. Ibid.

14. Ibid.

15. Ibid.

16. Church, in *Charleston City Yearbook*, 1929, xxvii.

17. In the heady economic times of the 1920s, the bridge promoters do not appear to have had trouble courting interested financiers. Information regarding the early financing of the Cooper River Bridge can be found in the James Allan Papers (1913–1930), South Carolina Historical Society, Charleston, S.C.

18. Grace, "A Review of the Cooper River Bridge."

**Chapter 4**

1. H. F. Church, "Chronological History of the Great Cooper River Bridge," *News and Courier*, August 9, 1929.

2. "Cooper River Bridge Hearing Today," *News and Courier*, September 26, 1927.

3. "Adverse Bridge Report Drafted," *News and Courier*, October 10, 1927.

4. Letter from J. J. Shiners to F. G. Davies, October 12, 1927, John P. Grace Papers, Special Collections Library, Duke University, Durham, N.C.

5. Letter from John P. Grace to J. J. Shiners, January 25, 1928, John P. Grace Papers, Special Collections Library, Duke University, Durham, N.C.

6. Grace, "A Review of the Cooper River Bridge."

7. At the time the Cooper River Bridge was being constructed, Waddell was involved in engineering rail projects for the Chinese government.

8. Letter from Egbert Hardesty to Jason Annan, April 5, 1999.

9. Grace, "A Review of the Cooper River Bridge."

10. Church, "Chronological History of the Great Cooper River Bridge."

11. Grace, "A Review of the Cooper River Bridge."

12. Ibid.

13. Charles K. Allen, "Depth of Mud and Water Offered Problems for the Bridge Engineers," *News and Courier*, August 8, 1929.

14. Ibid.

15. Though most sandhogs on the Cooper River Bridge project were African American, the term was industry jargon and was not pejorative.

16. "Sand Hogs Burrow 100 Feet under River to Build Span," *News and Courier*, August 8, 1929.

17. "Caisson Combined Unlucky Symbols," *News and Courier*, December 2, 1928.

18. "Soft Spot in Mud Bottom, Overload Concrete Blamed," *News and Courier*, December 2, 1928.

19. Correspondence pertaining to this lawsuit can be found in the John Patrick Grace Papers (1916–1940), South Carolina Historical Society, Charleston, S.C.

20. Joseph P. Riley to Pamela Gabriel, telephone conversation, May 2, 1996.

21. "The Story of the Bridge," pamphlet published by Francis Marion Hotel, 1929, Special Collections, College of Charleston, S.C.

22. Grace, "A Review of the Cooper River Bridge."

23. "The Story of the Bridge."

## Chapter 5

1. For advertisements and headlines, refer to the *News and Courier* issue for August 8, 1929.

2. Ibid.

3. Grace, quoted in "Let This Bridge Be the Emblem of Unity Declares John P. Grace, Its President," speech reprinted in *News and Courier,* August 8, 1929.

4. Jack Leland, "A Belle of a Bridge," *Post and Courier,* August 6, 1989.

5. John Hammond Moore, *The South Carolina Highway Department, 1917–1987* (Columbia: University of South Carolina Press, 1987), 148. Moore notes that a toll of fifty cents was typical for the several toll bridges operating in South Carolina during the 1920s.

6. See *News and Courier,* August 8, 1929.

7. "Given Outlet to the World," *News and Courier,* August 8, 1929.

8. The editorial from the January 1, 1930, *News and Courier* was reprinted by Mayor Thomas Stoney in *Charleston City Yearbook,* 1929, xxix.

9. Letter to the holders of Cooper River Bridge Corporation, from the Adjustment Committee for Cooper River Bridge Corporation, August 22, 1932, Dubose Heyward Papers, (1833–1972), South Carolina Historical Society, Charleston, S.C.

10. See the 1927 "Annual Report of the Cooper River Ferry Commission," in James Allan Papers (1913–1930), South Carolina Historical Society, Charleston, S.C.

11. The Cooper River Ferry Commission also received revenue from postal and freight fees in 1927. See 1927 "Annual Report of the Cooper River Ferry Commission," in James Allan Papers.

12. Moore, *South Carolina Highway Department,* notes that beginning in the 1920s the state highway department, in an effort to control all of the vital bridge links in the state highway system, had begun to buy the state's remaining toll bridges. The highway department did recommend that the state purchase the Cooper River Bridge (148).

13. Letter from Mayor Burnet R. Maybank to G. L. Buist Rivers, June 7, 1933, Cooper River Bridge Papers, Charleston City Archives, Charleston, S.C.

14. See the statements from state Senator John Dinkins in "Cooper River Bridge Purchase Approval Appears Assured," *News and Courier,* February 7, 1945.

15. "State Will Buy Cooper River Span, Lift Tolls in 1946," *News and Courier,* August 9, 1945.

16. See the bond statements issued by the Cooper River Bridge, Inc., Edward K. Pritchard Papers (1928–1950), South Carolina Historical Society, Charleston, S.C.

17. Maybank to Rivers, June 7, 1933, Cooper River Bridge Papers.

18. "State Will Buy Cooper River Span, Lift Tolls in 1946," *News and Courier,* August 9, 1945.

## Chapter 6

1. "Freeing of Bridge to Aid 'Forgotten County' of State," *News and Courier,* April 9, 1945.

2. Jesse R. Morillo, a crewman aboard the *Nicaragua Victory* in 1946, told his eyewitness account of the collision in an interview given to the *Post and Courier.* Terry Joyce, "Collector Recalls Drama of '46 Bridge Crash," *Post and Courier,* February 25, 1998.

3. Frank Tomich to the Charleston *News and Courier,* February 26, 1946.

4. "The Story of the Bridge."

5. In 1706, coastal Carolina was divided into ten divisions called parishes. Christ Church Parish, located east of Charleston, is bounded by the Wando River, the Atlantic Ocean, and Awendaw Creek.

6. U.S. Bureau of the Census, *1940 US Census Population of South Carolina* (Washington, D.C.: U.S. Government Printing Office, 1940), 4.

7. Though the increase appears high when compared to the 1929 population of 1,400 persons, the 1920 population of Mount Pleasant was about 1,600, indicating that the population of the town had remained consistent over the twenty-year period between 1920 and 1940.

8. Dale Rosengarten, "Bulrush Is Silver, Sweetgrass Is Gold," *Folklife Annual*, 1989, p. 154.

9. Mount Pleasant was indeed a "young" town: by comparison, in the entire greater Charleston area children under the age of fourteen comprised 32 percent of the population in 1950, and 36 percent of the population in 1960. See U.S. Census Bureau, "South Carolina, Population General Characteristics," *U.S. Census of Population 1950* (Washington, D.C.: U.S. Government Printing Office, 1951), 40.44, 40.65; U.S. Census Bureau, "South Carolina, Population General Characteristics," *U.S. Census of Population 1960* (Washington, D.C.: U.S. Government Printing Office, 1961), 42.32, 42.43.

10. The population of the Town of Sullivan's Island in 1950 was 895; its 1960 population was 1,358. Fort Moultrie, located on Sullivan's Island, was decommissioned in 1947. See U.S. Bureau of the Census, "South Carolina Number of Inhabitants," *U.S. Census of Population 1960* (Washington, D.C.: U.S. Government Printing Office, 1960) 42.10.

11. William Oliver Stevens, *Charleston* (New York: Dodd, Mead and Co., 1939).

12. Letter, Curtis Carter to Pamela Gabriel, July 16, 2001.

13. Some of the original sand dunes were reportedly thirty feet high. Due to archaic building restrictions the dunes could be leveled only by large scoops drawn by mules. J. C. Long sought, and was granted, permission to interpret the regulations to allow the use of bulldozers.

14. Interviews between Carmen Bunch and Pamela Gabriel, July 2001.

15. U.S. Bureau of the Census, "South Carolina Number of Inhabitants," *U.S. Census of Population 1970* (Washington, D.C.: U.S. Government Printing Office, 1971), 42.16.

16. Editorial, *Charleston Evening Post*, June 16, 1950.

17. Interview between James Craven and Jason Annan, April 28, 2002.

18. Response to a letter to the editor, *News and Courier*, March 21, 1952.

19. Otis Perkins, "The Big Roller Coaster," *News and Courier*, September 15, 1957.

20. Letter from Alex Sanders to Pamela Gabriel, February 29, 1996.

21. Letter from Claire Robinson to Pamela Gabriel, 1996.

22. Perkins, "The Big Roller Coaster."

23. See *News and Courier*, July 28, 1957.

24. Many of the new residents of Mount Pleasant settled in unincorporated areas immediately outside of the village center. The population of Christ Church Parish, the census district that covered most of the East Cooper area, including Mount Pleasant, grew from 5,165 persons in 1940 to 9,225 persons in 1950. In 1960, the census division of Christ Church Parish was split

between Mount Pleasant and McClellanville, and the town of Mount Pleasant annexed many unincorporated areas, boosting its population count.

25. "In-depth Inspection and Load Rating Analysis of the John P. Grace Memorial Bridge (US17-SBL) and the Silas N. Pearman Bridge (US17-NBL)—Charleston County," prepared by Sverdrup Corporation for the South Carolina Department of Transportation (SCDOT file no. 10-137A, 1987).

26. Maintenance costs continued to increase as the Grace Bridge aged. The cost to paint the bridge in 1984–85 was more than $2 million (based on a cost sheet suppled by the South Carolina Department of Transportation). Some of the factors contributing to an increase in the cost of maintaining the bridge are an increase in vehicle weight and speed, and changes in the salinity and speed of river currents. A report issued in 1992 stated that between 1961 and 1992, $13.4 million had been expended to maintain the Grace Bridge. Also, from 1978 until the issuance of the report, $18.3 million was spent on the Pearman Bridge, with an additional $15 million contracted for future painting and repair projects ("Replacement of the Cooper River Bridges on US17 over the Cooper River and Town Creek, Charleston County, S.C.," Environmental Impact Statement FHWA-SC-EIS-92-01-D, table, 1.17).

27. "Replacement of the Cooper River Bridges on US17," 1.18.

28. Laurie Fedon, "Cooper River Bridge—Facts at 50," *New and Courier*, July 29, 1979.

29. Pat Conroy, *Beach Music* (New York: Doubleday, 1995), 3.

30. Conroy, *The Prince of Tides* (Boston: Houghton Mifflin, 1986), 567.

31. Dorothea Benton Frank, *Sullivan's Island* (New York: Jove, 1999), 168.

32. As massive as the Ravenel Bridge will be, it will pale in comparison to the five-mile-long Øresund Bridge across the sound between Sweden and Denmark. The crossing in its entirety is ten miles long and consists of a cable-stayed span, a tunnel, and an artificial island; it took five years to complete, at a cost of $3 billion.

**Appendix**

1. These words are inscribed upon John Grace's tombstone.

2. For a full account of this murder read *Charleston Murders*, edited by Beatrice St. Julien Ravenel (New York: Duell, Sloan and Pearce, 1947).

3. Ibid, 188.

4. Ibid.

5. Boggs, "John Patrick Grace and the Politics of Reform in South Carolina, 1900–1931," 80.

6. Ibid.

7. "John P. Grace, City's Mayor for Eight Years, Succumbs," *News and Courier*, June 26, 1940.

# Bibliography

Allan, James, Papers, 1913–1930, South Carolina Historical Society, Charleston, S.C.

Boggs, Doyle W., Jr. "John Patrick Grace and the Politics of Reform in South Carolina, 1900–1931." Ph.D. diss., University of South Carolina, 1977.

*Charleston (S.C.) Evening News,* issues from 1929

*Charleston (S.C.) Evening Post,* issues 1910–1989

*Charleston (S.C.) News and Courier,* issues 1913–1989

*Charleston (S.C.) Post and Courier* issues 1989–present

*City of Charleston Yearbook* for years 1911, 1929, and 1930

Coleman, Richard. "Isle of Palms." *Sandlapper,* July 1969, pp. 41–47.

Conroy, Pat. *The Prince of Tides.* Boston: Houghton Mifflin, 1986.

Cooper River Bridge Papers, Charleston City Archives, Charleston, S.C.

Curtis, Elizabeth G. *Gateways and Doorways of Charleston, South Carolina, in the Eighteenth and the Nineteenth Centuries.* New York: Architectural Book Publishing Co., Inc., 1926.

Doyle, Don H. *New Men, New Cities, New South: Atlanta, Nashville, Charleston, Mobile, 1860–1910.* Chapel Hill: University of North Carolina Press, 1990.

Duffy, John. "Charleston Politics in the Progressive Era." Ph.D. diss., University of South Carolina, 1963.

Edgar, Walter. *South Carolina: A History.* Columbia: University of South Carolina Press, 1998.

Frank, Dorothea Benton. *Sullivan's Island.* New York: Jove, 1999.

Fraser, Walter, Jr. *Charleston! Charleston! The History of a Southern City.* Columbia: University of South Carolina Press, 1991.

Goldberg, David J. *Discontented America: The United States in the 1920s.* Baltimore, Md.: John Hopkins University Press, 1999.

Grace, John P. "A Review of the Cooper River Bridge." In "Cooper River Bridge Celebration, Charleston, S.C., August 8, 1929." Souvenir publication available at the Charleston *Post and Courier* Library, Charleston, S.C.

Grace, John Patrick, Papers, Special Collections Library, Duke University, Durham, N.C.

Grace, John Patrick, Papers, 1916–1940, South Carolina Historical Society, Charleston, S.C.

Heyward, Dubose, Papers, 1833–1972, South Carolina Historical Society, Charleston, S.C.

McIver, Petronia. *Names in South Carolina,* vol. 13. Columbia: Department of English of the University of South Carolina, 1966.

Moore, John Hammond. *The South Carolina Highway Department, 1917–1987.* Columbia, S.C.: University of South Carolina Press, 1987.

Pritchard, Edward Kriegsmann, Papers, 1928–1950, South Carolina Historical Society, Charleston, S.C.

Ravenel, Beatrice St. Julien, ed. *Charleston Murders.* New York: Duell, Sloan, and Pearce, 1947.

Rosen, Robert. *A Short History of Charleston.* Columbia: University of South Carolina Press, 1997.

Sparkman, Mary. *Through a Turnstile into Yesteryear.* Charleston, S.C.: Walker, Evans and Cogswell, 1966.

Stevens, William Oliver. *Charleston.* New York: Dodd, Mead and Company, 1939.

Sverdrup Corporation. "Report on the In-Depth Inspection and Safety Rating of the John P. Grace Memorial Bridge, Prepared for the South Carolina Highway Department." SCDOT file no. 10-137A, 1987, July 1988.

"The Story of the Bridge." Souvenir pamphlet published by Francis Marion Hotel, 1929. Special Collections, College of Charleston, Charleston, S.C.

U.S. Bureau of the Census. Published census data from 1900–1990.

Verner, Elizabeth O'Neill. *Mellowed By Time.* Charleston, S.C.: Bostick and Thornley, 1959.

White, Henry Alexander. *The Making of South Carolina.* New York: Silver, Burdett and Company, 1914.

# INDEX

Alhambra Hall, 103, 111

Allen, Charles Keyes, 58, 61, 63, 75, 77, 99

Allen, Charles Robinson, 30, 33, 34, 41, 42, 43

Ansonborough, 35

Armstrong, James, 79, 82

Army Corps of Engineers, 34, 101

Arthur Ravenel Jr. Bridge, 121, 122–23, 125

Ashley River, 1, 4, 30, 98; early bridges, 1, 2, 3, 4; bridge tollhouse, 2

Ashley River Memorial Bridge, 3, 4, 5, 40, 89

Atlantic Beach Hotel, 16, 17

Atlanticville, 14

Bailey span, 101, 102

Barkerding, Harry, 30, 33, 34, 41, 42, 43

basketweavers, 105

Bayonne Bridge, 50

Baytree, 81

Bayview Acres, 103

Beach Company, 106

Beauregard, General Pierre G. T., 3, 112

Ben Sawyer Bridge, 19, 107

bends. *See* caisson disease

Blease, Coleman L., 79

"blind tigers," 24

Bourbons, 37

Brawley, William, 43, 128, 130

Breach Inlet, 6, 8, 11; bridges over, 107

bridges, 46, 47, 48, 49, 50. *See also* listings under specific bridge names

Brooklyn Bridge, 77

Bunch, Carmen, 106–7

Butler, A. G., 79

Cadillac Investment Company, 40

caisson disease, 62

cantilevered truss, 50, 52, 53–57, 75, 111; anchor arm, 55, 56, 57, 67, 68, 71; cantilever arm, 55, 56, 57, 59, 67; suspended span, 55, 56, 57, 59. *See also* piers

Carquinez Strait Bridge, 75

Carr, C. D., 6

Castle Pinckney, 34

Charles E. Hillyer Company, 57

Charleston: City Hall, 37, 40, 128; economy of, xvi, 13, 22–24, 26, 39, 93, 95, 96; geographic location of, 1, 2, 4, 6, 8, 12, 26; growth of, xv–xviii, 4, 24, 26, 31, 89, 93, 98, 99; history of, 3, 6, 11, 13, 14, 16, 22; politics in, 3, 36–40, 93, 121; preservation of, 28, public health and infrastructure of, 6, 8, 18; 14, 22, 23, 26, 27, 40, 127, 128; shipping industry in, 17, 22, 30, 33, 128; tourism, 25, 26, 28, 32, 40; vices and crimes in, 24, 38, 40, 99, 129

Charleston Airport, 80, 86

*Charleston American*, 39, 128

Charleston and Seashore Railroad, 6, 8, 11

Charleston City High School, 35, 39

*Charleston City Yearbook*, 37, 127

Charleston Consolidated Railway Company. *See* Consolidated Railroad Gas and Electric

Charleston County, xvii, 3, 30, 33, 41, 48, 94, 95, 103, 105

Charleston County Sanitary and Drainage Commission, 33

Charleston Harbor, xv, xvi, 2, 4, 6, 10, 12, 17, 20, 21, 31, 34, 43, 52, 78, 102, 107, 125

Charleston Hotel, 28

Charleston Isle of Palms Traction Company, 16

Charleston Naval Base, 17, 23, 24, 39

Chisolm, John, 14
Christ Church Parish, 4, 6, 103, 105
Christian Brothers Academy, 35
City Railway Company, 11
Civil War, 1, 2, 14, 22, 35, 37, 103
Coastal Highway. *See* U.S. Highway 17
cofferdam, 59, 60
College of Charleston, 35, 39, 109
*Commodore*, 11
Coney Island, 4, 6
Conroy, Pat, 120
Consolidated Gas and Electric. *See* Consolidated Railroad Gas and Electric
Consolidated Railroad Gas and Electric, 8, 9, 11, 12, 16
Cooper River, 4, 6, 11, 17, 18, 30, 34, 40, 43, 50, 52, 53, 55, 58, 60, 71, 72, 74, 75, 77, 92, 95, 97, 99, 102, 113, 121
Cooper River Bridge: as a symbol of Charleston and source of memories, xv, xvi, 106, 107, 108, 109, 111, 120, 123, 125; construction of, 57–60, 62–77; construction accidents, 62, 63–65, 67; design of, 46, 48, 50–57; deterioration of, 118–19, 125–26; early planning of, xvi, 33–35, 42–43, 44–46, 48; effects on growth, xv, 104, 105–8, 118, 123, 125; financial troubles, xvi, 89, 91–93; franchise for, 30, 33, 34, 42; opening of, xvi, 78–87; sale and removal of tolls, xvi, 93–97, 98; ship collision, 99–102; replacement of, 108, 110, 111–13, 118, 120–23, 125. *See also* Cooper River Bridge Run
Cooper River Bridge Corporation, 28, 30, 31, 33, 34, 41, 43, 44, 46, 48, 50, 52, 63, 78, 89, 91, 94, 97
Cooper River Bridge Run, 113–17
Cooper River Ferry Commission, 7, 18–20, 21, 91–93
Cosgrove, John, 37, 63, 65
Cove Inlet, 6, 30; bridge, 7, 8, 18, 19, 107
Crescent Heights, 89

Davies, Frederick, 44
Devereaux, John Henry, 14
Don Holt Bridge, 121
Drum Island, 46, 48, 57, 68, 74, 75, 111

Eads, James, 60, 62
East Cooper, 4, 18, 19, 89, 98, 101, 105, 106, 108, 121, 122. *See also* Mount Pleasant
East Cooper Promotion Association, 111
East Shore Terminal Company, 41, 128
electric trolley. *See* trolley service
Elliot, William, 36, 37
*Evening Post*, 38, 72

Federal Securities, 43
ferry service, xv, 2, 3, 4, 6, 7, 8, 9, 11, 16, 17, 31, 42, 81, 92, 101, 102, 105. *See also* Cooper River Ferry Commission
Firth of Forth Bridge, 54, 75
Francis Marion Hotel, 28, 29, 40
Frank, Dorothea Benton, 120
Fricken, John, 3, 4
Fort Moultrie, 8, 13, 14, 15, 19, 79, 106, 111
Fort Sumter Hotel, 26, 27, 40
Foundation Company of New York, 57, 60, 62, 65

Gaillard, J. Palmer Jr., 111
Georgetown, S.C., 4, 5, 79, 83, 95, 96
Golden Gate Bridge, xv, 50, 54, 67
Grace, John Patrick, 1, 3, 4, 17, 24, 26, 34, 35–43, 46, 50, 61, 63, 78, 79, 81, 82, 89, 95, 102, 127–30
Great Depression, 33, 89, 90–92, 94, 105

*H. L. Hunley*, 107
H. M. Byllesby Company, 43, 46
Hallman, Harry, 121
Halsey, Alfred, 93
Hardaway Contracting Company, 106

Hardesty, Shortridge, 46, 47–48, 50, 52, 53, 75
Hell Gate Bridge, 50
Heyward, Dubose, 28
Hibben, Andrew, 6, 103
Highway Commission. *See* South Carolina Highway Commission
Hilliard, Oliver, 6, 103
Hobcaw Point, 6, 102
Hodges, Jim, 121
Hog Island, 6, 34,
Hollings, Ernest "Fritz," 121
Holton, Thomas, 6
Hort, William, 103
Hotel Seashore, 10, 11
*Housatonic*, 107
Hurricane Hugo, 29, 119
Hyde, Tristam T., 38, 39

immigration station, 24, 25
Imperial University, 47
I'on, Jacob Bond, 14
Isle of Palms, 4–12, 16, 19, 20, 21, 28, 31, 40, 78, 79, 80, 81, 88, 89, 95, 97, 101; tourism attractions, 10, 11, 13, 20, 106, 107, 113; modern growth of, 106–8, 113
Isle of Palms Beach and Racquet Club. *See* Wild Dunes Beach and Racquet Club

John P. Grace Memorial Bridge. *See* Cooper River Bridge
Johnston, Olin, 93
Jones, C. E., 79
Judge Mitchell King Mansion, 35
Jugnot, Charles, 6, 103

Kings Highway. *See* U.S. Highway 17

*Lawrence*, 20
Lawrence, Joseph, 6, 8, 11, 16
Lawson Family, 99, 100

Legare, William, 79
Leland, Jack, 80–81
Lempriere, Clement, 6
Lofton, James, 96
Logan, W. Turner, 37, 130
Long, J. C., 106, 107, 108, 130
Lords Proprietors, 6, 102
Lucas, Jonathan, 103

Manning, Richard, 24
Marion Square Park, 28, 29
Mark Clark Expressway, 121
Maybank, Burnett, 93, 94
McCants Ferry Company, 8
McCants, T. G., 88
McCants, William, 8
McClellanville, 4, 5, 79, 83, 89, 96
McClintic-Marshall Company, 57, 72
Medical University of South Carolina, 113
Memminger Home and School Association, 40
Miller, C. A., 43
Mills, Robert, 13
Milwaukee, Wis., 42
Morrison, H. T., 89
Moseley, Furman C., 85
Motte, Jacob, 103
Moultrie High School, 108, 109
Moultrie, William, 13
Moultrieville, 13, 14, 15
Mount Pleasant, 2, 4, 6, 7, 8, 9, 18, 20, 30, 31, 34, 74, 79, 81, 83, 88, 89, 92, 95, 99, 102–3, 111, 117; modern growth of, 98, 105–6, 107, 108, 110, 113, 118
Mount Pleasant and Sullivan's Island Ferry Service, 8

*Nansemond*, 18, 93
Naval Base. *See* Charleston Naval Base
New Brighton Hotel, 14, 16
New York City, 6, 22, 26, 36, 42, 46, 63, 130
Newberry, Marcus, 113

*News and Courier*, 4, 24, 26, 33, 45, 63, 72, 77, 78, 88, 89, 98, 108, 109, 119
*Nicaragua Victory*, 99–102

O'Sullivan, Capt. Florence, 12, 103

*Palmetto*, 20
Paramount Studios, 77, 79, 82
Parker, James L., 3
People's Building, 26, 27
Peterkin, Julia, 26
piers, 53, 58, 59, 63, 72, 74; anchor piers, 55, 56, 57, 62, 63, 74; cantilever piers, 55, 62, 74. *See also* Cooper River Bridge, design of; Cooper River Bridge, construction of
Pitt Street Bridge. *See* Cove Inlet Bridge
pneumatic caisson, 60, 62–65,
Poe, Edgar Allan, 14
Pohl, William G., 43, 91
Ports Utility Commission, 41, 44, 128
*Post and Courier*, 80
prohibition, 81, 129

Quebec Bridge, 50, 67, 75
Queensboro Bridge, 50, 75

railroads, xvii, 2, 6, 11, 22, 67, 72, 77, 128. *See also* listings under specific railroad names
Rensselaer Polytechnic Institute, 47
Revolutionary War, 6, 8, 12
Rhett, R. Goodwyn, 37
Richards, John, 42, 79
Riley, Joseph P., 40, 41, 69, 119, 121
Riverland Terrace, 89
Roosevelt, Theodore, 23
Rutledge, Archibald Hamilton, 81

Sanders, Alex, 109
sandhogs, 60, 62, 63, 65
Sanford, Mark, 113

Santee River Bridge, 3, 4, 5
*Sappho*, 20
Savannah, 4
Sciotoville Bridge, 53
Scott, Jonathan, 103
Sea Pines Company, 113
Seashore Railroad. *See* Charleston and Seashore Railroad
Sewee Indians, 6, 102
Shem Creek, 17, 103; bridges over, 18, 81, 105
Shinners, J. J., 43, 46
Silas N. Pearman Bridge, xv, 60, 111–13, 119, 125. *See also* Cooper River Bridge Run
slavery, 13, 103
Society for the Preservation of Old Dwellings, 28
Sottile, James, 16, 17, 18, 19, 20, 30, 40, 44, 48
South Carolina Department of Transportation. *See* South Carolina Highway Commission
South Carolina Highway Commission, 33, 34, 42, 109, 111, 122
South Carolina Highway Department. *See* South Carolina Highway Commission
South Carolina Interstate and West Indian Exposition, 22
Southern Renaissance, 26
St. Lawrence Cemetery, 43
St. Paul's Episcopal Church, 13
Stella Maris Roman Catholic Church, 14
Stevenson, Thomas, 72
Stoney, Thomas, 4, 31, 40, 52, 78, 79, 81, 82
suburbs, 4, 98, 102, 105–6, 125. *See also* Mount Pleasant; East Cooper; West Ashley
Sullivan, Frank, 41, 42, 46
Sullivan's Island, 4, 6, 7, 8, 9, 11, 12–16, 17, 19, 81, 106, 107

T. Allen Legare Jr. Bridge, 4

Taft, William Howard, 24

The Crescent, 4

The Groves, 81

Tillman, Benjamin, 17, 130

tolls, 3, 4, 19, 20, 33, 81, 88, 89, 95–97, 111

tourism, 25–29, 31, 45, 88, 89, 90, 91, 92, 94, 101. *See also* Isle of Palms, tourist attractions; Charleston, tourism

Town Creek, 50, 53, 55, 59, 68, 69, 72, 74, 75, 77

trains. *See* railroads

trolley service, 6, 7, 8, 9, 17, 20

truss, 53, 55, 74. *See also* cantilevered truss

Union Pier, 44, 45

U.S. Army Corps of Engineers. *See* Army Corps of Engineers

U.S. Customs House, 45

U.S. Highway 17, 2, 4, 5, 79, 84, 86, 105, 109, 110

Virginia Bridge and Iron Company, 57, 67

Waddell and Hardesty, 46

Waddell, John A. L., 46–47, 48, 61

Wallace, Oliver T., 95, 96

Wando Indians, 102

Wando River, 102

Wappetaw, 102

Wappoo Hall, 89

Wappoo Heights, 4, 89

War Between the States. *See* Civil War

War Department, 18, 19, 44, 45, 46

Washington Race Course, 22

West Ashley, 4, 89

White Point Gardens, 26, 27, 40, 117

Wild Dunes Beach and Racquet Club, 113

Williams, Ransome, 95

Wilson, Woodrow, 39

Windermere, 89, 98

Wolfe, Thomas, 26

Woodland Shores, 89

World's Fair, 10, 11

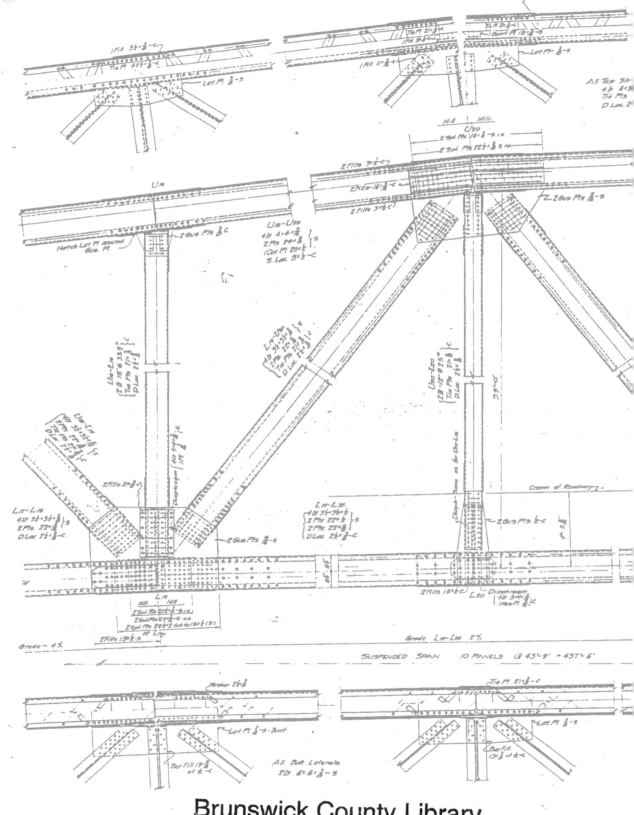